Student Exoplanet Projects Using Data from the Kepler Mission

Online at: https://doi.org/10.1088/2514-3433/ad7c3e

AAS Editor in Chief

Ethan Vishniac, Johns Hopkins University, Maryland, USA

About the program:

AAS-IOP Astronomy ebooks is the official book program of the American Astronomical Society (AAS) and aims to share in depth the most fascinating areas of astronomy, astrophysics, solar physics and planetary science. The program includes publications in the following topics:

GALAXIES AND
COSMOLOGY

INTERSTELLAR
MATTER AND THE
LOCAL UNIVERSE

STARS AND
STELLAR PHYSICS

EDUCATION,
OUTREACH,
AND HERITAGE

HIGH-ENERGY
PHENOMENA AND
FUNDAMENTAL
PHYSICS

THE SUN AND
THE HELIOSPHERE

THE SOLAR SYSTEM,
EXOPLANETS,
AND ASTROBIOLOGY

LABORATORY
ASTROPHYSICS,
INSTRUMENTATION,
SOFTWARE, AND DATA

Books in the program range in level from short introductory texts on fast-moving areas, graduate and upper-level undergraduate textbooks, research monographs, and practical handbooks.

For a complete list of published and forthcoming titles, please visit iopscience.org/books/aas.

About the American Astronomical Society

The American Astronomical Society (aas.org), established 1899, is the major organization of professional astronomers in North America. The membership (~7,000) also includes physicists, mathematicians, geologists, engineers, and others whose research interests lie within the broad spectrum of subjects now comprising the contemporary astronomical sciences. The mission of the Society is to enhance and share humanity's scientific understanding of the universe.

Student Exoplanet Projects Using Data from the Kepler Mission

Michael C LoPresto

Department of Astronomy, University of Michigan, Ann Arbor, MI, USA

IOP Publishing, Bristol, UK

ISBN   978-0-7503-6290-0 (ebook)
ISBN   978-0-7503-6288-7 (print)
ISBN   978-0-7503-6291-7 (myPrint)
ISBN   978-0-7503-6289-4 (mobi)

DOI   10.1088/2514-3433/ad7c3e

Supplementary material is available for this book from https://doi.org/10.1088/978-0-7503-6290-0.

Version: 20241101

AAS–IOP Astronomy
ISSN 2514-3433 (online)
ISSN 2515-141X (print)

British Library Cataloguing-in-Publication Data: A catalogue record for this book is available from the British Library.

Published by IOP Publishing, wholly owned by The Institute of Physics, London

IOP Publishing, No.2 The Distillery, Glassfields, Avon Street, Bristol, BS2 0GR, UK

US Office: IOP Publishing, Inc., 190 North Independence Mall West, Suite 601, Philadelphia, PA 19106, USA

For my wife Jan, and children Sarah (Anthony), Emily and Sam

Contents

Preface

This project had its origin in activities I developed for students doing directed studies in conjunction with introductory astronomy courses to work with exoplanet data taken by the Kepler mission. I wrote articles about several of the activities that were published in the journal *Physics Education* (IOP) and *The Physics Teacher* (AAPT). I was then approached by editors from the IOP about basing a book on the articles that instructors could use to have their students do similar studies, which is what we have here.

Chapter 1 is an introduction to the Kepler mission and its goals, then Chapters 2–5 are the procedures and results of studies done with Kepler mission data obtained online. Chapters 2 and 3 are studies about planetary habitability and Chapters 4 and 5 are about what can be learned from analyzing the densities of the exoplanets when both mass and radius data is available. Each chapter includes not only the results of the studies, but comments on the procedures and pedagogy used with the students. At the end of each chapter are suggested projects that instructors can assign to students including reproducing the studies discussed, as well as more guided activities that can be found in the appendices along with the data. Chapter 6 is about ongoing and future missions that will continue the study of exoplanets and the search for habitability and life that was begun by the Kepler mission.

Acknowledgements

A sincere thank you to the students who chose to do astronomy directed studies with me for which these activities were originally developed and the IOP editors who first noticed my work and reached out to me about writing this book, as well as those who reviewed and suggested how to improve my proposal.

Michael C. LoPresto,
Ann Arbor, MI 2024

Author biography

Michael C. LoPresto

Michael C. LoPresto is currently a *Lecturer* in the Department of Astronomy at the University of Michigan-Ann Arbor. He teachers a range of introductory general-education astronomy courses, largely to non-science majors, several of which include exoplanets, their possible habitability and life in the Universe.

Dr. LoPresto has been teaching introductory astronomy since 1990 and has published nearly 100 journal articles including many on astronomy education and has written a conceptual Introductory astronomy textbook, a textbook supplement and several laboratory manuals.

He received a BS in Physics from Edinboro University of Pennsylvania in 1987, an MS in Physics and MS in Physics Education from the University of Michigan in 1989, and Eastern Michigan University in 1996 and his PhD in astronomy from James Cook University in 2012. He taught astronomy and physics at Henry Ford College in Dearborn, Michigan from 1990 to 2017 and after doing a sabbatical and postdoc at the University of Michigan focused on astronomy education, he returned there in 2017 as a full-time lecturer.

Mike lives in Saline, Michigan with his wife of 35 years Jan, with adult children Sarah (Anthony), Emily and Sam all close by. He is a dedicated daily exerciser, an avid reader of science, science fiction and fantasy, a *Star Trek* fanatic, and a life-long fan of University of Michigan-Wolverines football—*GO BLUE*!

Introduction

An independent or directed study done within the confines of a semester by a student in conjunction with an introductory general-education astronomy course can simply be a library-researched paper and/or presentation in which a student explores a topic that they find interesting in greater detail than will be covered by the course. However, a more interesting and perhaps more useful project can be to guide the student through an investigation that is more similar to actual scientific research. Through such a project, the student can still learn more about the topic that interests them and also be exposed to the scientific process. This is a perspective often lacking in introductory general education science courses.

To provide such an opportunity, the instructor needs to have ready an interesting and engaging activity with an initial goal that can be reached by guiding a student through straightforward analysis of provided or easily attainable data from which reasonable conclusions can be reached within a semester. This is what is offered in the coming chapters of this book. The data, procedures and results of directed studies done with students of introductory astronomy on a subject at the forefront of astronomical research, extra solar planets.

Student Exoplanet Projects Using Data from the Kepler Mission

Michael C LoPresto

Chapter 1

The Kepler Mission

This chapter contains a brief overview of the Kepler mission, how the data was obtained for the studies detailed in coming chapters and some demographics based on the data; definitions of the different exoplanet types and a comparison of the numbers of each different type discovered by the mission.

1.1 Introduction

NASA's Kepler Mission (https://www.jpl.nasa.gov/missions/kepler) operated from 2009 until 2018 with a primary goal of determining whether or not potentially habitable planets, planets similar in size to Earth and orbiting within their stars' *habitable zones*, the range of distances at which temperatures allow water to exist as a liquid, are common in our galaxy.

1.2 The Kepler Mission

Figure 1.1 is an artist's conception of the Kepler space telescope that made use of the *transit method* of exoplanet detection (Figure 1.2). This is a measurement of the apparent decrease in the star's brightness measured by a device known as a *photometer* when a planet passes in front of its star. The photometers used for the mission were extremely sensitive. They could detect changes as small as 0.01% in the star's apparent brightness. The percentage decrease in the star's brightness is proportional to the ratio of the square of the planet's radius to the square of star's known radius.

The amount of time between repetitions of identical transits is the *orbital period*, p, of the planet, from which the average orbital radius, a, or semi-major axis of the planet's elliptical orbit could be determined using Kepler's third law of planetary motion $p^2 = a^3$.

After taking data on nearly 200,000 stars in a specific target region (Figure 1.3) of the Milky Way, the Kepler mission had discovered over 2000 planets orbiting F, G, K and M spectral-type stars. Determination of which if any are potentially

doi:10.1088/2514-3433/ad7c3ech1

Figure 1.1. The Kepler space telescope. (Courtesy of NASA images, http://www.nasa.gov.)

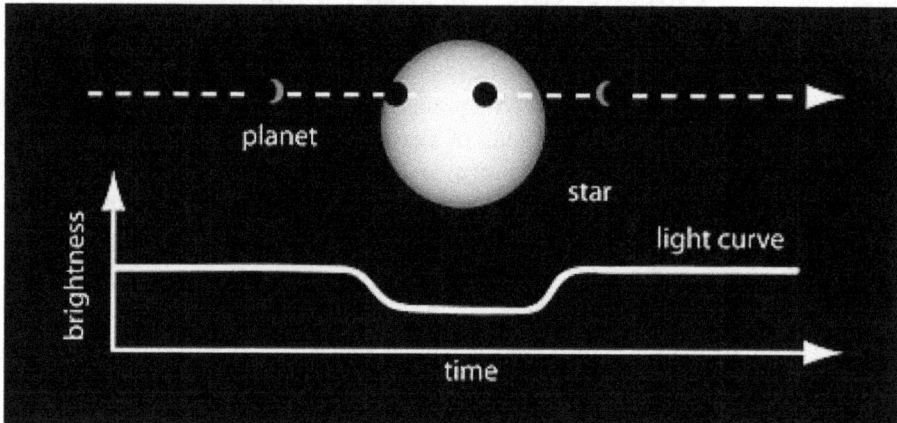

Figure 1.2. The *transit method* of exoplanet detection. (Courtesy of NASA images, http://www.nasa.gov.)

habitable, the goal of the study described in Chapter 2, could be a first step in answering one of the most fundamental questions ever asked by the human species, whether or not life and possibly intelligence, exists elsewhere in our galaxy.

1.3 Obtaining the Data

Data from the Kepler mission (and others) is available in the *NASA Exoplanet Archive* (https://exoplanetarchive.ipac.caltech.edu) shown in Figure 1.4.

Clicking on the **Planetary Systems** button shown in Figure 1.4 and entering "Kepler" in the space in the "Host Name" column, as shown in Figure 1.5, brought

Figure 1.3. The Kepler mission's target region. (Courtesy of NASA images, http://www.nasa.gov.)

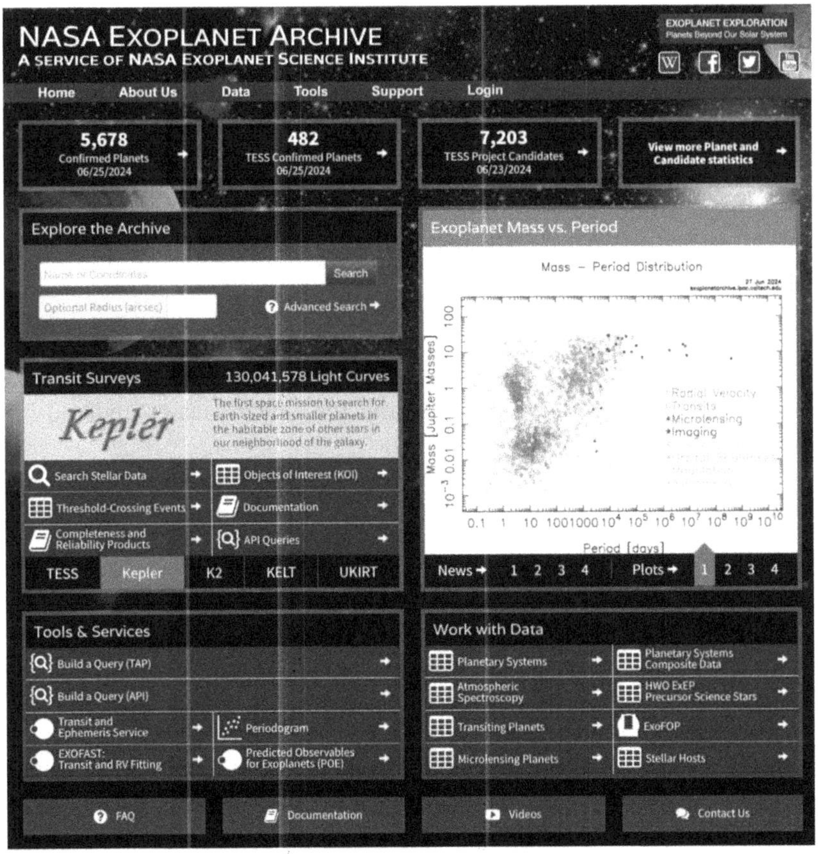

Figure 1.4. The NASA Exoplanet Archive. (Courtesy of NASA.)

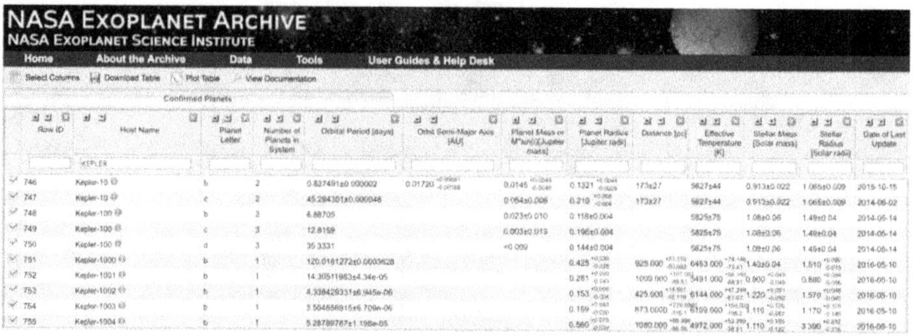

Figure 1.5. Display of data from Kepler mission in the *NASA Exoplanet Archive*. (Courtesy of NASA.)

	A	B	C	D	E	F	G	H	I	J	K	L
1	pl_hostname	pl_letter	pl_pnum	pl_orbper	pl_orbsmax	pl_bmassj	pl_radj	st_dist	st_teff	st_mass	st_rad	rowupdate
2	Kepler-10	b	2	0.837491	0.0172	0.0145	0.132	173	5627	0.91	1.06	10/15/15
3	Kepler-10	c	2	45.2943		0.054	0.21	173	5627	0.91	1.06	6/2/14
4	Kepler-100	b	3	6.88705		0.023	0.118		5825	1.08	1.49	5/14/14
5	Kepler-100	c	3	12.8159		0.003	0.196		5825	1.08	1.49	5/14/14
6	Kepler-100	d	3	35.3331		0.009	0.144		5825	1.08	1.49	5/14/14
7	Kepler-1000	b	1	120.0181			0.425	925	6453	1.4	1.51	5/10/16
8	Kepler-1001	b	1	14.30512			0.281	1009	5491	0.9	0.88	5/10/16
9	Kepler-1002	b	1	4.336429			0.153	425	6144	1.22	1.57	5/10/16
10	Kepler-1003	b	1	3.554857			0.159	873	6109	1.11	1.17	5/10/16

Figure 1.6. Downloaded data from Kepler mission transferred to an Excel spreadsheet.

up a wealth of data on over 2000 exoplanets discovered by the Kepler mission. Data was downloaded from the *NASA Exoplanet Archive* in spreadsheet form (CSV format was selected) using the "Select Columns" and "Download Table" menus seen in Figure 1.5.

Figure 1.6 shows spreadsheet data for the first few of over 2000 planets. The orbital periods (column D) were given in units of days, the orbital semi-major axes (E) in AU, the planet masses. and radii (F and G) in "Jupiters," the stellar distances and temperatures in parsecs and Kelvins and the stellar masses and radii in solar units (the values for the Sun = 1).

This data set, available in Appendix A (or at this book's IOP homepage at http://iopscience.iop.org/mono/978-0-7503-6290-0) is what was used for the studies outlined in Chapters 2 and 3.

1.4 Exoplanet Types—Demographics

The different exoplanet types shown in Figure 1.7 as originally defined by the NASA Kepler mission are classified by size. Exoplanets that are: 1.25 Earth radii or less are considered terrestrial or "*Earths*"; those between 1.25 and 2 Earth radii are called "*Super-Earths*," a new type of planet discovered by the Kepler mission of which there is *not* an example in our solar system; 2–6 Earth radii are considered "*Neptunes*," distinguishing them from the gas-giant "*Jupiters*," 6–15 Earth radii; and "*Larger*," >5 Earth radii exoplanets. There is also an "overlap" category

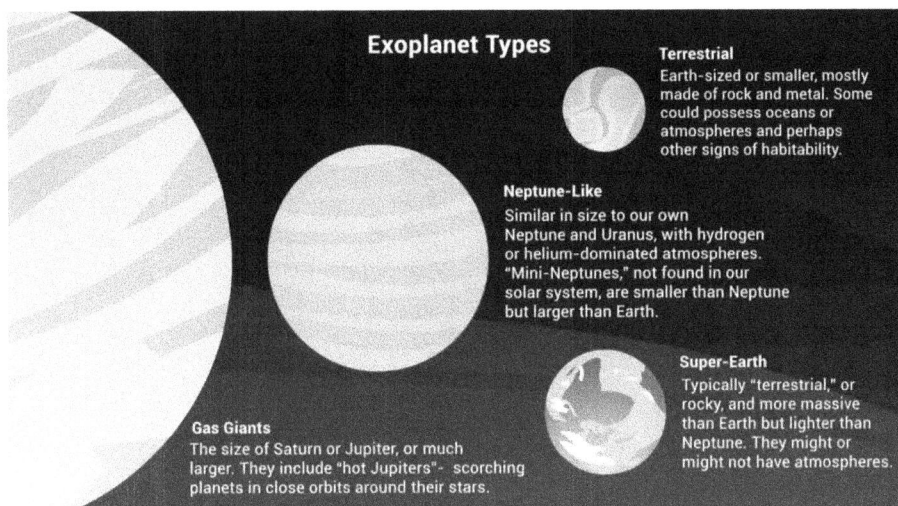

Figure 1.7. Different types of exoplanets defined by the NASA Kepler mission. Yaqoob (2011), credit: NASA/ JPL-Caltech/Lizbeth B. De La Torre.

Table 1.1. Numbers of Each Planet Type Discovered by the Kepler Mission

Planet Type	Radius (Earths)	Number	% of $N = 2276$
Earth-like	<1.25	333	14.6
Super-earth	1.25–2	725	31.9
1 Neptune	2–6	1075	47.2
2 (Ice Giant)			
3 Jupiter	6–15	127	5.6
4 (Gas Giant)			
Larger	>15	15	0.7

between larger *Super-Earths* and smaller *Neptunes*, up to 3 or 4 Earth radii, that are considered "*Mini-Neptunes*," of which there is also no example in our solar system. Chapter 5 describes a study attempting to distinguish Mini-Neptunes from Neptunes and Super-Earths.

Table 1.1 was compiled from the data set. Figure 1.8 histogram comparing the numbers of each exoplanet type discovered by the Kepler mission.

1.5 Conclusion

The data obtained from the *NASA Exoplanet Archive* (available in Appendix A or at this book's IOP homepage) was used in instructor-directed student studies that will be described in Chapters 2 and 3. The first is on potential habitability of exoplanets, based on their distance from their star and the stars' habitable zones, the second on estimates of their temperatures.

Number of Exoplanet Type

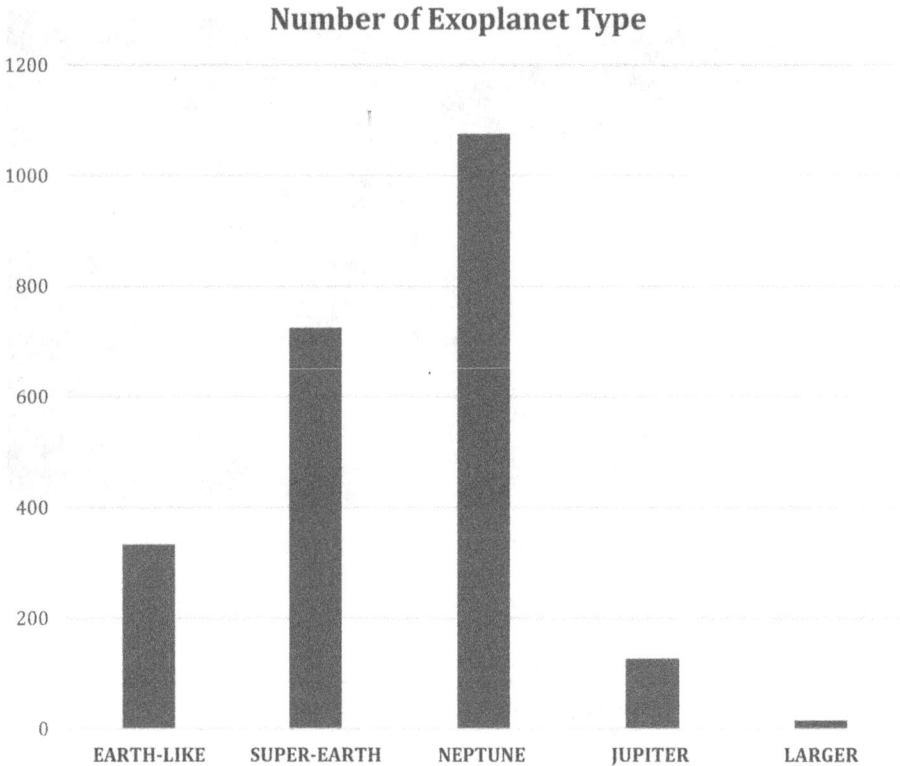

Figure 1.8. Plot of the number of each planet type discovered by the Kepler mission.

The study presented in Chapter 4 are about using the data to compare the densities of a subset of exoplanets discovered by the Kepler mission for which the masses and radii are known (Appendix D or at this book's IOP homepage, http://iopscience.iop.org/mono/978-0-7503-6290-0)) to the densities of planets of our own solar system of similar size-classification. The study in Chapter 5 is an attempt to use calculated densities to distinguish between Mini-Neptunes, Neptunes and Super-Earths.

Student Projects and Assignments

1 Assign reading of:
   Exoplanet Exploration—Planets Beyond Our Solar System https://exoplanets.nasa.gov/what-is-an-exoplanet/planet-types/overview/ (or a similar reference) for a student to become familiar with exoplanet detection methods and the different types of exoplanets.
2 Assign a student to make use of the data in Appendix A (or at this book's IOP homepage, http://iopscience.iop.org/mono/978-0-7503-6290-0) to produce a table and histogram similar to Table 1.1 and Figure 1.8.

References and Further Reading

Exoplanet Exploration—Planets Beyond Our Solar System https://exoplanets.nasa.gov/what-is-an-exoplanet/planet-types/overview/

Gould, A., Komatsu, T., DeVore, E., Harman, P., & Koch, D. 2015, PhTea, 53, 201

Johnson, J. A. 2016, How Do You Find an Exoplanet (Princeton, NJ: Princeton Univ. Press)

LoPresto, M. C., & Ochoa, H. 2017, PhysEd, 52, 065016

Yaqoob, T. 2011, Exoplanets and Alien Solar Systems (Baltimore, MD: New Earth Labs-Education and Outreach)

Chapter 2

Earth-sized Planets in Habitable Zones

The *habitable zone* of a star is the range of distances from it where temperatures allow the existence of water in its liquid state on the surface of a planet with sufficient atmospheric pressure. This chapter is an overview of a study meant to identify potentially habitable exoplanets using actual data available online from NASA's Kepler mission.

2.1 Background

The expected temperature, T_p, of a planet of radius, R_p, an average orbital radius or distance, d, from a star can be estimated based on the fraction of the radiation emitted by a star of Temperature, T_s, and radius, R_s, that the planet absorbs and reemits. This is done with the Stephan–Boltzmann law, $P = \sigma\, T^4$ ($\sigma = 5.67 \times 10^{-8}$ W m^{-2} × K^4 is the Stefan–Boltzmann constant). Setting the fraction of the power emitted by the star that a planet absorbs can be set equal to the power emitted by the planet giving

$$\sigma 4\pi R_s d^2 \otimes T_s{}^4 \times (\pi R_p{}^2/4\pi d^2) = \sigma 4\pi R_p d^2 T_p{}^4. \tag{2.1}$$

The quantity in parenthesis ($\pi R_p{}^2/4\pi d^2$) is the fraction of the power radiated by the star over a spherical area, $4\pi d^2$, that is absorbed by the cross-sectional area, $\pi R_p{}^2$, of the planet at the planet's orbital distance, d. After rearranging and cancellations (2.1) simplifies to give

$$T = T_S \times (R_S/2d)^{0.5}. \tag{2.2}$$

The boundaries of our Sun's habitable zone have been estimated to be as close as 0.5 AU and as far 3 AU. Using the Sun's temperature of 5800 K and radius of 6.95×10^8 m and 1 AU $= 1.5 \times 10^{11}$ m, these distances correspond to temperatures calculated with (2.2) of 395 K and 161 K. This range of distances includes Venus, 0.72 AU, and Mars, 1.52 AU, and is considered a very optimistic habitable zone. Solving (2.2) for d and using the range of temperatures only at which water can exists as a liquid, $T = 373$ K to $T = 273$ K, a necessity for life, gives a more pessimistic

doi:10.1088/2514-3433/ad7c3ech2   2-1

habitable zone of about 0.56 AU to 1.04 AU. This still includes Venus, but not Mars, and barely includes Earth!

The temperature of a planet does *not* depend only on radiation absorbed from its star. It also depends on the reflectivity or *albedo* of its surfaces and the amount *greenhouse effect* occurring in its atmosphere. Albedo is the percentage of radiation incident on a planet's surfaces that is reflected. Reflection can lower a planet's surface temperature allowing it to be closer to a star than expected from radiation alone, while still maintaining a temperature high enough for water to be liquid. A quantitative treatment of this will be seen in Chapter 3.

Greenhouse warming occurs when gases in a planet's atmosphere such as water vapor, carbon dioxide and other "greenhouse gases" absorb a portion of the energy from the star absorbed and reemitted by the planet. This could allow a planet to still be habitable at a further distance from a star than expected only due to radiation. There will also be a quantitative treatment of this in Chapter 3.

Due to the possible cooling of a planet due to albedo and warming due the greenhouse effect, the habitable zone for a star could be wider than a range calculated with (2.1) for only radiation and the temperatures for liquid water. Considering this and also the optimistic and pessimistic limits for habitable zones discussed above, reasonable estimates for the habitable zone for our Sun could be from about 0.6 AU to 2 AU. The inner estimate is just outside the 0.56 AU, calculated above with (2.1), limited for liquid water and includes Venus, the outer estimate includes Mars. This suggests that slightly different circumstances could have led to planets in their orbits being potentially habitable. Perhaps a less massive planet with a thinner atmosphere experiencing less greenhouse warming than occurred on Venus could have allowed a planet at or near 0.72 AU to be cooler and possibly habitable. A more massive planet than Mars at 1.52 AU could have retained a thicker atmosphere and experienced more greenhouse warming, resulting in more habitable temperatures.

Using (2.2) at 0.6 AU and 2 AU gives estimates of the inner and outer (radiation) temperatures of our Sun's habitable zone of 360 K and 197 K. These temperatures can now each be used as T in Equation (2.3) below, which is (2.2), solved for d, with meters converted to AU (1 AU = 1.496×10^{11} m) and the units of R_S to solar radii ($R_{Sun} = 6.96 \times 10^8$ m). This allows estimation of the limits of the habitable zone of any star in AU based on its Kelvin temperature and its radius in units of solar radii;

$$d = 0.00232 \text{ AU} \times (R_S/R_{Sun}) \times (T_S/T)^2. \qquad (2.3)$$

Once these limits are calculated for a star, they can be compared to the orbital radii of planets in orbit to determine whether or not they lie within a habitable zone, once again recalling that a planet's albedo could allow it to be closer to its star and greenhouse effect could allow it to be farther than what would be calculated based strictly on radiation from the star and the freezing and boiling points of water.'

2.2 Procedure and Pedagogy

Once a student's interest in doing a directed study is established, a weekly meeting time can be set up. The exoplanets planets chapter of their textbook should be assigned as reading with a focus on the transit detection method (Figure 1.2) which should then be discussed in detail at the next meeting.

Whether the instructor should do the derivations of (2.2) and (2.3) or simply introduce the equations, or have the student attempt it, as suggested in **Student Projects and Assignments 1 and 2** below, depends on the mathematical level of the course and mathematics and physics background of the student. For a student in a general education course, it may be best to simply introduce (2.1) and do the derivations of (2.2) and (2.3) for the student.

At this point, the spreadsheet containing the data set (available in Appendix A or at this book's IOP homepage, http://iopscience.iop.org/mono/978-0-7503-6290-0) can then be introduced. The next assignment should be to use the spreadsheet to convert, using information found in Table 1.1, a textbook or online, the mass and radius data from units of "Jupiters" to "Earths" and generate column for each in "Earth = 1" units. If the student's facility manipulating data and doing calculations with spreads sheets is a problem, time should be spent on developing those skills. Assigning too many tasks or steps between meetings can result in confusion, causing some tasks to be done incorrectly and have to be repeated. A comfortable pace to avoid such confusion that should still allow a project to be completed within a semester is only assigning one (1) task after each meeting.

After the data is converted as indicated above, reading about different exoplanet types should be assigned. Although many references are available, the one given in **Student Projects and Assignments 1** in Chapter 1 (https://exoplanets.nasa.gov/what-is-an-exoplanet/planet-types/overview/) is recommended.

2.3 Analyzing the Data

First, the data set should be sorted on the planet radii (column-G as shown in Figure 1.6) in increasing order. Planets any larger than Super-Earths are likely composed mostly gas and liquid like the Jovian planets of our solar system while it is considered likely that Super-Earths are composed of rock and metal as are our solar system's Earth-like, terrestrial planets. So, to search for potentially habitable planets, data for all planets other than *Earth-like* and *Super-Earth* should be deleted (their radii ranges are found in Table 1.1). To leave a margin for error, planets of a radii of up to 2.3 Earths, 15% above the Super-Earth cut-off could be retained.

The spreadsheet data can then be sorted by increasing star temperature (column-I in Figure 1.6). Figure 2.1 is a comparison of the numbers of each spectral type of star found by the Kepler mission to have Earth-like planets or Super-Earths in orbit. Table 2.1 shows this data, the temperature and mass range for each star type and the inner-most and outer-most limits of the habitable zones calculated with (2.3).

A data set for each spectral type of star in Table 2.1 should now be extracted and sorted in order of increasing distance of the planets from their stars (column E in Figure 1.6). Planets for which there is no orbital distance given in the data have to be discarded. This will reduce the total number of planets making the data more manageable.

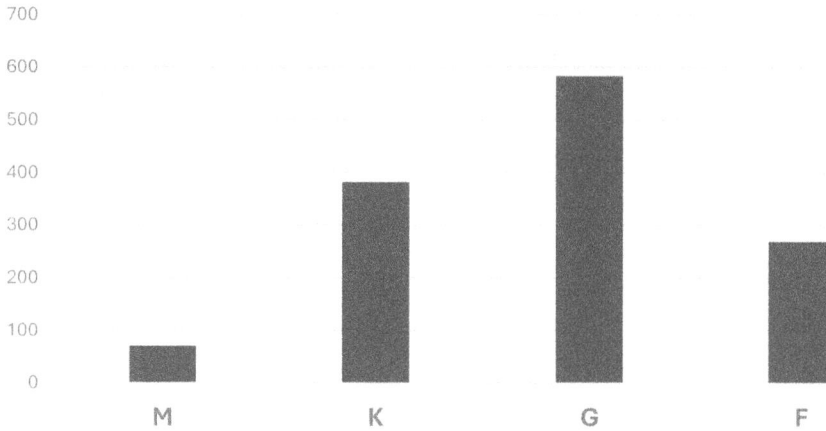

NUMBER of Stars of Spectral Type

Figure 2.1. Plot of the numbers of M, K, G and F-type stars surveyed by the Kepler mission.

Now, the inner and outer limits of the habitable zone of every star remaining can

Table 2.1. The Temperatures, Masses, and Range of Habitable Zone Limits of M, K, G and F-type Stars Surveyed by the Kepler Mission.

Star Type	Temperature Range	Mass Range	INNER HZ	OUTER HZ	Number of Stars	Percentage of Stars
M-(red) dwarfs	<4000 K	<.5 Sun	0.03 AU	0.5 AU	70	5.4%
K-dwarfs	3900 K–5200 K	0.45–0.8 Sun	0.2 AU	1.4 AU	382	29.3%
G-Sun-like	5300 K–6000 K	0.8–1.2 Sun	0.45 AU	2.5 AU	583	44.7%
F-hotter	>6000 K	1–1.4 Sun	0.75 AU	5.2 AU	269	20.6%

be calculated using the spreadsheet with (2.3) and compared to the orbital radius of their exoplanet(s). The seventeen (17) exoplanets, shown in Table. 2.2, about 0.04% of the over 2000 discovered by the Kepler mission were found to lie within the habitable zones of the M, K and G-type stars in the data set. Note that *no* planets were found in the habitable zones of F-type stars.

2.4 Results

A student can now be sent to *Habitable Exoplanets Catalog* maintained by the *Planetary Habitability Laboratory* at the University of Puerto Rico at Arecibo http://phl.upr.edu/projects/habitable-exoplanets-catalog to confirm their results.

The exoplanets in Table. 2.2 are all listed in the *Habitable Exoplanets Catalog* and are considered to be the most promising potentially habitable planets discovered by the Kepler mission. This group is often referred to as "*Kepler's Hall of Fame*" (Figure 2.2).

Table 2.2. The 17 Planets Discovered by the Kepler Mission that Orbit within Their Star's Estimated Habitable Zones.

pl_hostname	pl_letter	pl_pnum	pl_orbper	pl_orbsmax	pl_bmassE	pl_radE	st_teff	st_mass	st_rad	i_hz	o_hz	t-planet
Kepler-236	c	2	24.0	0.132	0	1.9936	3750	0.56	0.51	0.128	0.429	198
Kepler-438	b	1	35.2	0.166	0	1.12	3748	0.54	0.52	0.131	0.437	224
Kepler-296	e	5	34.1	0.169	0	1.5232	3740	0.5	0.48	0.120	0.401	225
Kepler-296	f	5	63.3	0.255	0	1.8032	3740	0.5	0.48	0.120	0.401	247
Kepler-186	f	5	129.9	0.432	0	1.1648	3755	0.54	0.52	0.131	0.438	257
Kepler-235	e	4	46.2	0.213	0	2.2176	4255	0.59	0.55	0.178	0.595	271
Kepler-440	b	1	101.1	0.242	0	1.904	4134	0.57	0.56	0.171	0.572	286
Kepler-155	c	2	52.7	0.242	0	2.24	4508	0.58	0.62	0.226	0.753	290
Kepler-437	b	1	66.7	0.288	0	2.128	4551	0.71	0.68	0.252	0.842	303
Kepler-283	c	2	92.7	0.341	0	1.8144	4351		0.57	0.193	0.645	304
Kepler-442	b	1	112.3	0.409	0	1.344	4402	0.61	0.6	0.208	0.695	320
Kepler-62	e	5	122.4	0.427	35.934	1.6128	4925	0.69	0.64	0.278	0.928	325
Kepler-174	d	3	247.4	0.677	0	2.184	4880		0.62	0.264	0.883	327
Kepler-62	f	5	267.3	0.718	34.98	1.4112	4925	0.69	0.64	0.278	0.928	329
Kepler-439	b	1	178.1	0.563	0	2.24	5431	0.88	0.87	0.459	1.534	337
Kepler-69	c	2	242.5	0.64	0	1.7136	5638	0.81	0.93	0.529	1.767	348
Kepler-452	b	1	384.8	1.046	0	1.624	5757	1.04	1.11	0.659	2.199	355

The rows highlighted in red are for planets orbiting M-type stars, orange, K-type and yellow, G-type. Note the thinner and closer habitable zones of the cooler M-type stars compared to the wider and farther ones of the hotter K-type and still hotter G-type stars.

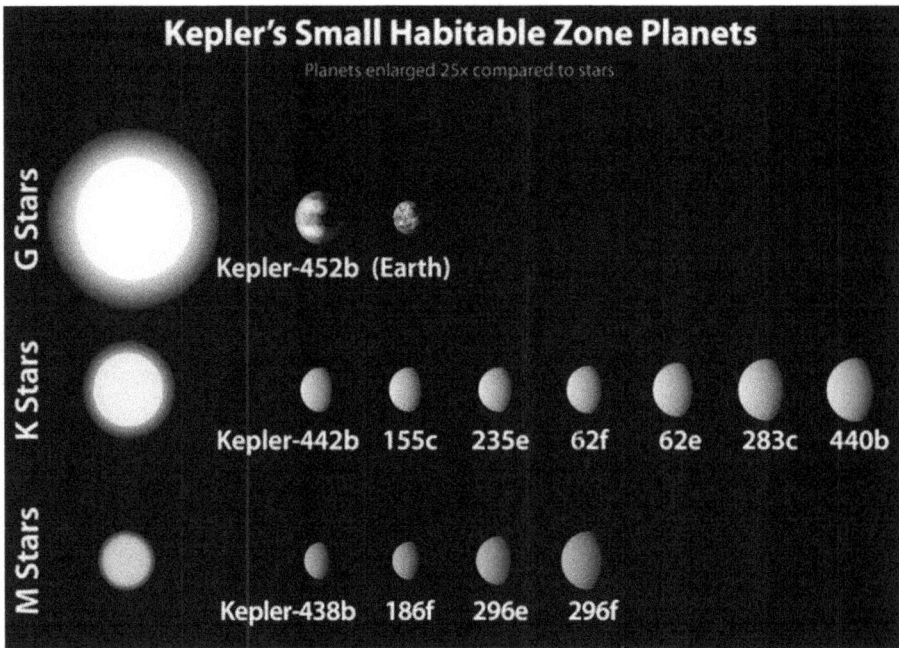

Figure 2.2. Members of the *Planetary Habitability Laboratory's* list of potentially habitable exoplanets discovered by the Kepler mission. (Courtesy of NASA images, http://www.nasa.gov.)

Planets in Habitable Zones of M-stars

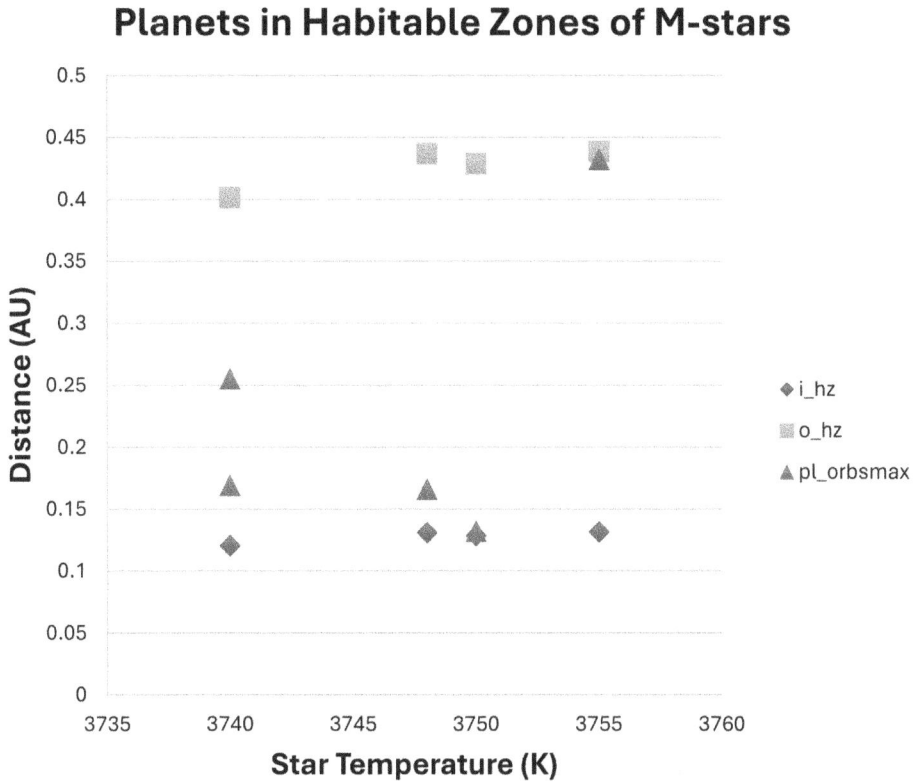

Figure 2.3. Planets discovered by the Kepler mission with orbital distances within the limits of the habitable zones of M-type stars.

Plots like Figures 2.3 and 2.4 can also be made. They are plots of the orbital radii of the planets listed in Table 2.2 compared inner and outer habitable zones of their stars. Again, note that the habitable zones of the cooler M-type stars are closer to the star and thinner, while those of the hotter K-type and M-type stars are farther and wider.

Figure 2.5 shows Kepler-186f orbiting its M-type star in a much closer habitable zone than that of Kepler-452 or our Sun, both G-type stars. Kepler-452b, dubbed Earth 2.0, when discovered in 2015 may be the exoplanet most similar to Earth yet known. It has a radius 1.6 times that of Earth's (Figure 2.6) and orbits its star at a distance just over 1 AU.

2.5 Conclusion

A study similar to that outlined above can be done by an undergraduate student supervised by a faculty member in a semester. The project can provide valuable experience for the student by demonstrating several important aspects involved in the process of scientific research including the analysis of a large data set including quantitative components. The results are also satisfying, being largely in agreement with those found in sources on exoplanets and their possible habitability.

Planets in Habitable Zones of K and G-stars

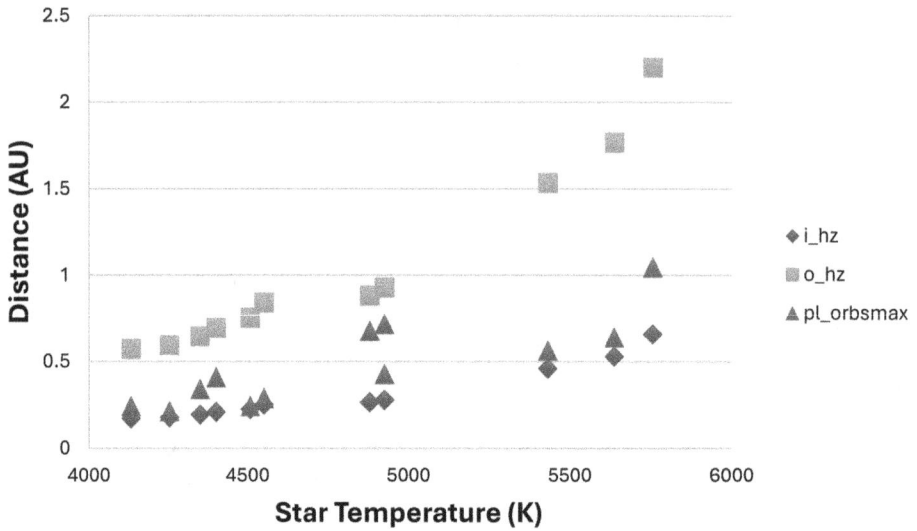

Figure 2.4. Planets discovered by the Kepler mission with orbital distances within the limits of the habitable zones of K (<5200 K) and G (>5200 K) stars.

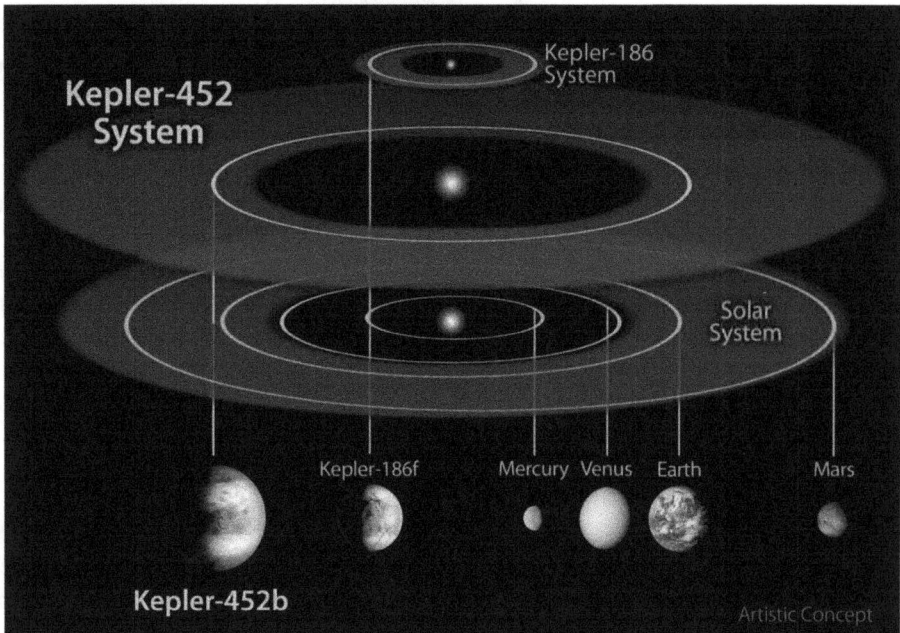

Figure 2.5. The habitable zones of the Kepler-186 and 452 systems, an M-type and G-type star, compared to that of our Sun (also a G-type star). (Courtesy of NASA images, http://www.nasa.gov.)

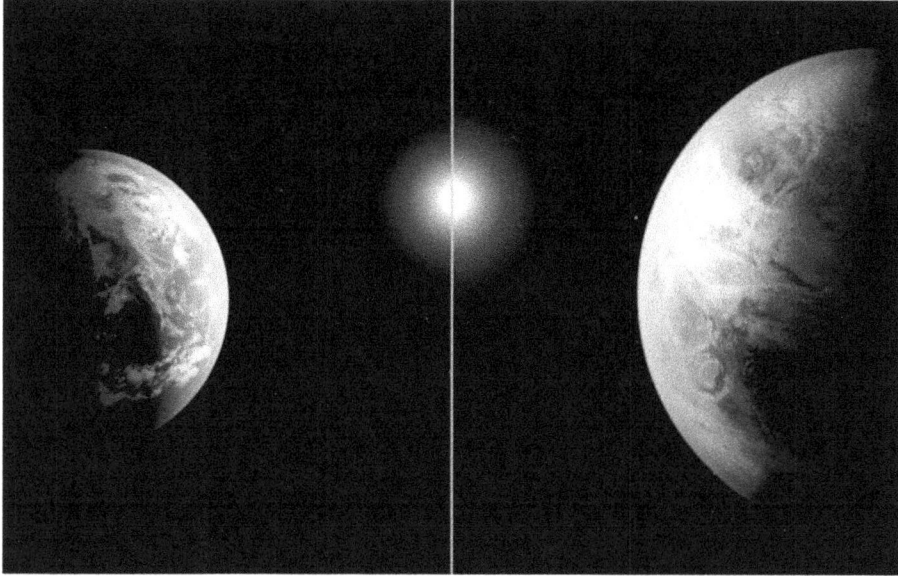

Figure 2.6. Artist's comparison of Earth (left) and Kepler-452b. (Courtesy of NASA images, http://www.nasa.gov.)

Student Projects and Assignments

1a-Assign a student the problem of deriving Equation (2.2), the expected temperature, T_p, of a planet of an average orbital radius or distance, d, from the star a star of temperature, T_s, and radius, R_s, using the Stefan–Boltzmann Law, $P = \sigma T^4$, where $\sigma = 5.67 \times 10^{-8}$ W m^{-2} × K^4 is the Stefan–Boltzmann constant, by setting the fraction of the power emitted by the star that the planet absorbs equal to the power emitted by the planet.

    1b-Also assign the derivation of Equation (2.3) by solving (2.2) for d and using conversion factors 1 AU = 1.496×10^{11} m and $R_{Sun} = 6.96 \times 10^8$ m.

2 Making use of the data set provided in Appendix A (or at this book's IOP homepage, http://iopscience.iop.org/mono/978-0-7503-6290-0), engage a student in a directed study similar to that outlined in this chapter.

3 Appendix B provides a faster, guided, possibly group, activity for students to use the data to find the possibly habitable planets discovered by the Kepler mission.

4 Have a student research and report on the potential problems with the habitability of planets located within the habitable zones of M and/or F spectral-type stars.

References and Further Reading

Chou, F., & Johnson, M. 2015, M NASA's Kepler mission discovers bigger, older cousin to Earth, https://www.nasa.gov/press-release/nasa-kepler-mission-discovers-bigger-older-cousin-to-earth

Crockett, C. 2017, Observers caught these stars going supernova, Science News, https://www.sciencenews.org/article/observers-caught-these-stars-going-supernova

Livio, M., & Silk, J. 2017, PhT, 70, 50

LoPresto, M. C. 2024, Horizons in World PhysicsVol. 312, ed. A. Reimer (Hauppauge, NY: Nova Publishing)

LoPresto, M. C. 2019, PhTea, 57, 159

LoPresto, M. C., & Ochoa, H. 2017, PhysEd, 52, 065016

Chapter 3

Calculating the Temperatures of Possibly Habitable Planets

An alternative approach to determine if a planet is possibly habitable is to use the data to estimate its temperature and see if it falls within the boundary temperatures of a habitable zone, the freezing, 273 K, and boiling, 373 K, temperatures of water. Estimates of the effects of the planet's reflectivity or albedo on the temperature as well as the effect of greenhouse warming can also be considered.

3.1 Introduction "Radiation" Temperatures

Derived in Chapter 2, Equation (2.2) is an expression for the temperature of a planet, T_p, as a function of its orbital distance, d, from its star in terms of the radius, R_s, and temperature, T_s, of the star,

$$T_p = T_{S*}(R_S/2d)^{0.5}. \tag{3.1}$$

Using the values for radius of the Sun and the AU in meters (3.1) can be converted to an expression for a planet's temperature in terms of the radius of its star in solar radii and the orbital distance of the planet, in AU,

$$T_p = 0.048 \ Ts * [(R_s/R_{\text{Sun}})/d)]^{0.5}. \tag{3.2}$$

This expression (3.2) can be used to calculate "radiation" temperatures. Exoplanet temperatures based only on consideration of radiation from their star.

3.2 Albedo and the Greenhouse Effect

As mentioned in Chapter 2, The surface temperature of a planet is *not* only dependent only on radiation from its star. It also depends on the reflectivity or *albedo* of its surfaces and the amount *greenhouse effect* occurring in its atmosphere.

Albedo is the percentage of radiation incident on planets' surfaces that they reflect. This can lower a planet's surface temperature allowing it to be closer to a star than expected from calculations considering radiation alone while still maintaining a temperature high enough for water to be liquid. For instance, Earth's average albedo is about $a = 0.31$, meaning that its surface on average reflects about 31% of the radiation incident upon it from the Sun. Due to radiation alone, Earth's average temperature calculated with (3.1) would be about $T_r = 279$ K.

An expression for how albedo affects a planet's temperature is

$$T_a = T^*(1 - a)^{0.25}, \tag{3.3}$$

Using this expression for Earth's $a = 0.31$, gives a temperature over 10% lower than the radiation temperature, $T_a = 249$ K.

Greenhouse warming occurs when gases in a planet's atmosphere such as water vapor, carbon dioxide and others absorb a portion of the incident radiation from a star after the planet absorbs and reemits it. This could allow a planet to still be habitable at a further distance from a star than expected only due to radiation (and albedo). An expression for the effect of greenhouse warming on the temperature of a planet is

$$T_g = T^*(1 + \tau)^{0.25}, \tag{3.4}$$

where τ is the "thickness" of an atmosphere, compared to a $\tau = 1$ thickness of Earth's atmosphere. Using (3.4) to calculate the greenhouse warming in Earth's atmosphere, raises the temperature calculated from radiation and albedo with (3.3) by almost 20% to about 296 K. In comparison, for Mars' thin atmosphere, $\tau = 0.2$, and for the extremely thick atmosphere of Venus $\tau = 92$!

3.3 Procedure and Pedagogy

Whether done as a follow up to the study in Chapter 2 or as a stand-alone study, this study is not as involved as the previous one. Whether to derive, simply introduce or have the student derive (3.1) and (3.2) possibly staring with (2.1), as suggested in **Student Projects and Assignments-1a and 1b** below would again depend on the mathematical level of the course and the mathematics and physics and background of the student. Despite being a less involved study, it is still good practice to limit the tasks assigned to one per meeting. The equations could be introduced and discussed first, then the data introduced at the next meeting with the assignment to calculate radiation temperatures only, after that albedo and the greenhouse effect can be discussed, perhaps in successive meetings, with successive assignments to use (3.3) and (3.4) to calculate temperatures modified by each. **Student Projects and Assignments-3**, that involves calculating the temperatures of the planets in our own solar system first with radiation, then modified by albedo and the greenhouse effect could be used as an effective prelude to this study.

3.4 Results

Table 3.1 includes exoplanets already determined in Chapter 2 (see Table 2.2) to be within their star's habitable zones. The second to last column is their "radiation"

Table 3.1. Temperatures of "Potentially Habitable" Planets Identified in Chapter 2 (Table 2.2) Calculated Considering Radiation Only with (3.2) Then Modified for the Net Effects of Albedo (3.3) and the Greenhouse Effect (3.4).

Host Name and Planet	Orbital Distance (AU)	Planet Radius (Earth = 1)	Star Temperature (K)	Star Type	Star Radius (Sun = 1)	Planet Temperature (K)- a	Planet Temperature (K) Adjusted-b
Kepler-155c	0.242	2.24	4508	K	0.62	348	378
Kepler-174d	0.677	2.184	4880	K	0.62	225	245
Kepler-186f	0.432	1.1648	3755	M	0.52	198	216
Kepler-235e	0.213	2.2176	4255	K	0.55	329	358
Kepler-236c	0.132	1.9936	3750	M	0.51	355	386
Kepler-283d	0.341	1.8144	4351	K	0.57	271	295
Kepler-296f	0.255	1.8032	3740	M	0.48	247	269
Kepler-296e	0.169	1.5232	3740	M	0.48	304	330
Kepler-437b	0.288	2.128	4551	K	0.68	337	366
Kepler-438b	0.166	1.12	3748	M	0.52	320	348
Kepler-439b	0.563	2.24	5431	G	0.87	325	354
Kepler-440b	0.242	1.904	4134	K	0.56	303	329
Kepler-441b	0.64	1.68	4340	K	0.55	194	211
Kepler-442b	0.409	1.344	4402	K	0.6	257	279
Kepler-452b	1.046	1.624	5757	G	1.11	286	311
Kepler-62f	0.718	1.4112	4925	K	0.64	224	244
Kepler-62e	0.427	1.6128	4925	K	0.64	290	316
Kepler-69c	0.64	1.7136	5638	G	0.93	327	356
Venus	0.72	0.95	5800	G	1	329	734-c
Earth	1	1	5800	G	1	279	304-c
Mars	1.52	0.53	5800	G	1	227	221-c

a "Radiation" Temperatures calculated with (3.2).

b Temperatures calculated also considering albedo (3.3) and greenhouse effect (3.4).

c Temperatures for planets in our solar system calculated with known albedos (Mars $a = 0.25$; Venus, $a = 0.75$) and atmospheric thicknesses (Mars, $\tau = 0.2$; Venus, $\tau = 92$). Students can be guided through these calculations in the activity in Appendix C).The data set is available in Appendix A (or at this book's IOP homepage, http://iopscience.iop.org/mono/978-0-7503-6290-0).

temperatures calculated with (3.2). Note that none are greater than 373 K, but some of the exoplanets, those near the outer edge of their stars' habitable zones, have temperatures less than 273 K. However, before any conclusions an attempt should be made to consider the effects of albedo and greenhouse warming.

Making a generalization that all these "potentially habitable" planets are also "Earth-like" and using Earth's values for albedo and atmospheric thickness, $a = 0.3$ and $\tau = 1$ results in a multiplicative factor from (3.3) of $(1 - 0.3)^{0.25}$ for albedo and $(1 + 1)^{0.25}$ for greenhouse warming from (3.4). So, $[(1 - 0.3)^{0.25}] * [(1 + 1)^{0.25}] = [0.7 * 2]^{0.25} = 1.4^{0.25} - 1.088$. The combination of Earth-like albedo and greenhouse warming causes an overall increase of nearly 9%. above "radiation temperatures." These temperatures are found in the last column of Table 3.1.

Even after being adjusted for albedo and greenhouse warming, temperatures of several of the exoplanets in Table 3.1, Kepler-174d, 186f, 441b, 62f still appear "too cold," to be potentially habitably, at about 10%–20% below 273 K, but they all have radii > 1.25 that of Earth, making them "Super-*Earths*." This could mean they are more massive than Earth and have thicker atmospheres and therefore possibly more greenhouse warming, so they could still be considered potentially habitable.

Kepler-174d, radius, $R = 2.184$ Earths, and possibly Kepler-296 (temperature just below 273 K), $R \sim 1.8$ Earths, could be Mini-Neptunes. Being composed largely of liquids and gases and thus they would be less likely to be habitable. *Mini-Neptunes* will be discussed in Chapter 5.

Exoplanets in Table 3.1 with adjusted temperatures up to about 4% above 373 K, Kepler-155c, 236c, may well be "too hot," as they also, all also being larger than Earth, could have stronger greenhouse effects resulting in even higher temperatures and are also large enough to possibly be considered *mini-Neptunes*.

If starting with this study rather than the one outlined in Chapter 2, not knowing which exoplanets are found within their stars' habitable zones, the approach could be to calculate the temperatures of all the exoplanets in the original data set, found in Appendix A (or at this book's IOP homepage, http://iopscience.iop.org/mono/978-0-7503-6290-0) then to investigate which fall into a habitable temperature range.

3.5 Conclusion

Estimating the temperatures of exoplanets provides an alternative to attempting to determine whether or not a planet lies in a star's habitable zone (as done in the study in Chapter 2) as a way to determine potential habitability. Temperatures can be estimated with radiation alone or, with the effects of reflectivity and greenhouse warming (based on estimates of a planet's albedo and atmospheric thickness). After exoplanet temperatures have been calculated, discussions, like those above, especially for planets at temperatures near the higher and lower limits of potential habitability, about whether the size of an exoplanet or other factors could affect its temperature should be encouraged.

Student Projects and Assignments

1a-Assign a student the problem of deriving Equation (3.1), the expected temperature, T_p, of a planet of an average orbital radius or distance, d, from the star a star of temperature, T_s, and radius, R_s from Equation (2.1).

1b-Also assign the conversion of Equation (3.1) to Equation (3.2) using the conversion factors 1 AU $= 1.496 \times 10^{11}$ m and $R_{Sun} = 6.96 \times 10^8$ m.

2-a Making use of the-data set provided in Appendix A, (or at this book's IOP homepage, http://iopscience.iop.org/mono/978-0-7503-6290-0) engage a student in a directed study similar to that outlined in this chapter

2-b If the study in Chapter 2 has been done, making use of the data set provided in Appendix A (or at this book's IOP homepage, http://iopscience.iop.org/mono/978-0-7503-6290-0) to estimate planetary temperatures as done in this study, determine if any planets not initially identified as potentially habitable have estimated temperature within the habitable range.

3-Appendix C is an activity in which a student is guided through calculating the temperatures of the planets of our own solar system; first with radiation from the Sun, then considering albedo (3.3) and the greenhouse effect (3.4).

4- Have a student explore how use of albedos in (3.3) and/or atmospheric thickness in (3.4) different than the "Earth-like" estimates made in this chapter affect the estimates of planetary temperatures. For example, perhaps a more massive planet than Earth would have a more substantial atmosphere and greater "thickness" value than $t = 1$, or one less massive a lesser value. Perhaps some planets may have different surface conditions and therefore different albedos.

References and Further Reading

https://en.wikipedia.org/wiki/Circumstellar_habitable_zone
LoPresto, M. C. 2019, PhyEd, 54, 013002
LoPresto, M. C. 2013, PhTea, 51, 161
LoPresto, M. C. 2018, PhTea, 56, 459
LoPresto, M. C. 2019, PhTea, 57, 568
LoPresto, M. C. 2024, *Horizons in World Physics* Vol. 312, ed. A. Reimer (Hauppauge, NY: Nova Publishing)
LoPresto, M. C. 2019, PhTea, 57, 159
LoPresto, M. C., & Hagoort, N. 2011, PhTea, 49, 113
LoPresto, M. C., & Ochoa, H. 2017, PhEd, 52, 065016

Chapter 4

Calculating the Densities of Extra Solar Planets

The over 2000 *exoplanets* discovered by the NASA Kepler mission were found mostly with the *transit method* of detection (Figure 1.2). The *transit method* provides the radius of the exoplanet. Among these over 2000 exoplanets, there are just over 200, about 10%, for which not only the radius is known, but also the mass. The masses were determined by *the Doppler detection method* (Figure 4.1). The mass and radius together allow for the calculation of an exoplanet's density. Knowledge of the density of an exoplanet allows for more direct comparison to the densities of similar size-classification planet types in our own solar system.

4.1 Introduction and Background

The main focus of the Kepler mission was the discovery of Earth-like planets and their possible habitability. The possibility of the discovery of life on exoplanets is very intriguing, but contrary to popular belief, largely because it is portrayed this way in the media, the search for habitable planets and extra-terrestrial life is *not* the entire focus of the study of exoplanets. This study and the one in Chapter 5 still make use of Kepler mission data, but are concerned with what can be learned from knowing the densities of exoplanets.

The planets of our solar system are considered to fit into one of two major groups: the smaller, lower mass, Earth-like or *terrestrial* planets of higher density due to their rocky-metal composition and the larger, higher mass, Jupiter-like, *Jovian* planets of lower density due to gas and liquid composition (more icy composition in the case of Uranus and Neptune). Table 4.1 contains the mass, radii, and density of our solar systems planets in terms of the values for Earth.

The data set available in Appendix D (or at this book's IOP homepage, http://iopscience.iop.org/mono/978-0-7503-6290-0) includes over 200 (234) of the over 2000 (2311) exoplanets discovered by the Kepler mission for which both radius *and* mass data are available. The radii were determined by the *transit* method, (see Figure 1.2), masses are determined with the *radial velocity* or *Doppler detection*

doi:10.1088/2514-3433/ad7c3ech4   4-1

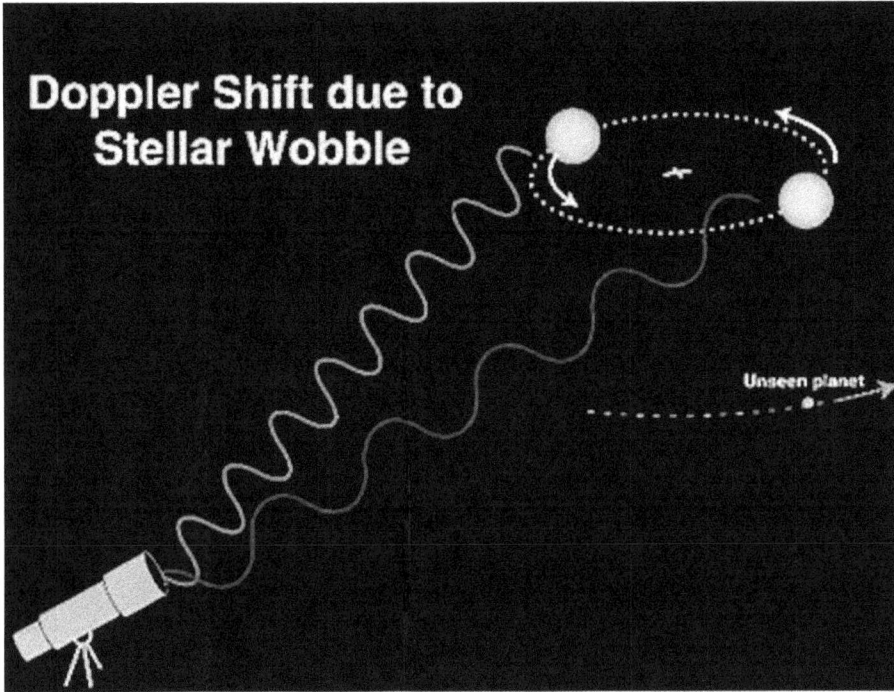

Figure 4.1. The *radial velocity* or Doppler detection method of exoplanet detection. (Reproduced from NASA, www.nasa.gov.)

Table 4.1. The Masses, Radii, and Densities of the Planets of Our Solar System (Earth Mass = Earth Radius = Earth Density = 1).

Planet	Mass (Earth = 1)	Radius (Earth = 1)	Density (Earth = 1)
Mercury	0.06	0.38	0.98
Venus	0.82	0.95	0.95
Earth	1.00	1.00	1.00
Mars	0.11	0.53	0.71
Jupiter	318.26	11.21	0.24
Saturn	95.14	9.45	0.13
Uranus	14.54	4.01	0.23
Neptune	17.09	3.88	0.30

method, shown in Figure 4.1. The radial velocity or Doppler detection method of exoplanet detection uses Doppler shifts that appear in the light coming from the star as both a star and an unseen exoplanet orbit the center of mass their system. This was in fact the first method used for exoplanet detection. These Doppler shifts vary in a regular pattern with a repetition period equal to the orbital period of the planet.

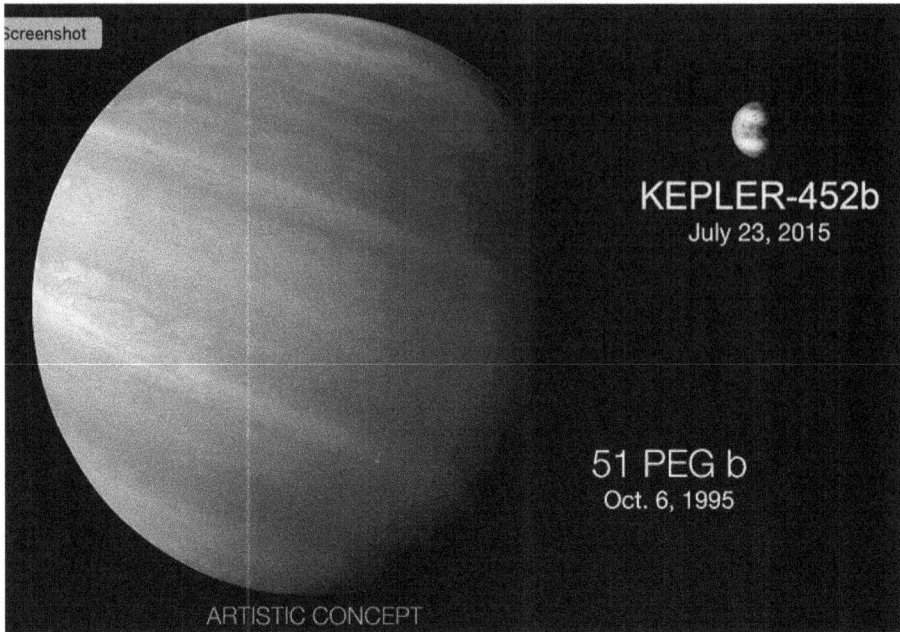

Figure 4.2. Artist's conception of 51Pegasi-b compared to the Earth-like Kepler 452-b. (Reproduced from NASA, www.nasa.gov.)

From this, the orbital radius is determined and is used with the known mass of the star to determine the mass of the planet.

The first exoplanet identified, was 51Pegasi-b https://science.nasa.gov/exoplanet-catalog/51-pegasi-b/ in 1995. The exoplanet is a *"Hot-Jupiter,"* an exoplanet type named for being a planet about half the mass of Jupiter that orbits so close to its Sun-like (at just over 0.05 AU) star that it has an orbital period of only about 4 days! The discoverers of 51Pegasi were awarded the 2019 Nobel Prize in physics. Figure 4.2 shows an artist's conception comparing 51Pegasi-b to Kepler 452-b (see Chapter 2) the exoplanet most similar to Earth yet discovered.

The *transit method* of exoplanet detection gives us the radius, r, of an exoplanet. The radial velocity or Doppler detection method gives us the mass, m, of a planet. Knowing both mass and radius allows us to calculate the density of an exoplanet. Density, $\rho = M/V$. M is the mass of an exoplanet and V is its volume. The volume, V, of a sphere or radius r is $V = 4/3 * \pi r^3$. Therefore, the density of an exoplanets of mass, m, and radius, r, is;

$$\rho = 3*m/(4*\pi r^3). \tag{4.1}$$

Density can be used to estimate whether an exoplanet is more likely a solid, rock and metal, planet, similar to our solar system's high-density, *terrestrial* planets or a gas and liquid or icy planet like our low-density, *Jovian* planets (see Table 1 for the densities of our solar system's planets). It is important to note. however, that an exoplanet could potentially be composed of many different combinations of materials so it should *not* be assumed that density alone can be used to determine the exact composition of an exoplanet. This study is simply meant to use their densities to determine whether exoplanets are *similar* in composition to those of our solar system of the same size-classifications.

4.2 Procedure and Pedagogy

As discussed for the study in Chapter 2, after time for a weekly meeting is set up, reading of the extra solar planets chapter of their textbook assigned this time focusing on both the transit and Doppler detection methods, both of which should be discussed in detail at the next meeting. Next the spreadsheet data set available in Appendix D (or at this book's IOP homepage, http://iopscience.iop.org/mono/978-0-7503-6290-0) can be introduced. This is a considerably smaller data set than the one from Appendix A used in Chapter 2. It consists of the names of the above-mentioned 234 planets, their orbital semi-major axes, and the mass and radii, again, as with the Appendix A data in Chapter 2 study both in units of "Jupiters." So again, the first assignment using the spreadsheet data set should be to convert, using information found in Table 4.1, a textbook or online the mass and radius data from units of "Jupiters" to "Earths." This will provide a first look at whether the student's skill level in manipulating and doing calculations with spreadsheets is a problem. If the student has trouble at this point, some time should be spent on developing those skills.

Also, as discussed in Chapter 2, assigning too many tasks or steps between meetings could result in confusion, causing some to be done incorrectly and have to be repeated. A comfortable pace to avoid such confusion that will still allow the project to be completed within a semester is only assigning one (1) task after each meeting. The next task should be to use the spreadsheet data set to generate a density column in "Earth = 1" units by dividing the converted mass column by the cube of the converted radius data column data.

After the data is prepared as indicated above, as seen in Student Projects and Assignments—1 below, reading about different exoplanet types should be assigned, https://exoplanets.nasa.gov/what-is-an-exoplanet/planet-types/overview/ *Exoplanet Exploration* -Planets Beyond Our Solar System is an excellent reference for this.

Now the task, of sorting the data by increasing radius can be assigned so the numbers of each type of exoplanet (by size category—see Figure 1.7 and Table 1.1) can be determined and tables and histograms like Table 4.2 and Figure 4.3 can be generated.

Generating the first plot, one like Figure 4.4, of log-density versus radius for the entire data set should then be assigned as the task for the next meeting. Plots of log-density (as in Figure 4.4) can be preferable to those of density, the log-plots are vertically more-compact, making them simpler to read and analyze. To do this, a new column in the spreadsheet for the LOG(DENSITY) will need to be created.

Before assigning separating the data and making plots for individual exoplanet size-categories, as seen in Student Projects and Assignments-2, reading of *Planet Classification*: How to Group Exoplanets https://www.space.com/36935-planet-

Table 4.2. The Number and Percentage of Each Exoplanet Type, as Defined by the Kepler Mission (Figures 1.7 and 1.8) for which Mass and Radius are Both Known (Available Online from *NASA's Kepler Mission* (https://exoplanetarchive.ipac.caltech. edu)—Available in Appendix D (or at This Book's IOP Homepage http://iopscience.iop. org/mono/978-0-7503-6290-0).

Planet Type	Radius Range (Earth = 1)	Number	Percentage %
Earth-LIKE	<1.25	22	9.4
	1.25–2	41	17.5
Neptunes	2–6	109	46.6
Jupiters	6–15	52	22.2
Larger	>15	10	4.3
Total		234	100

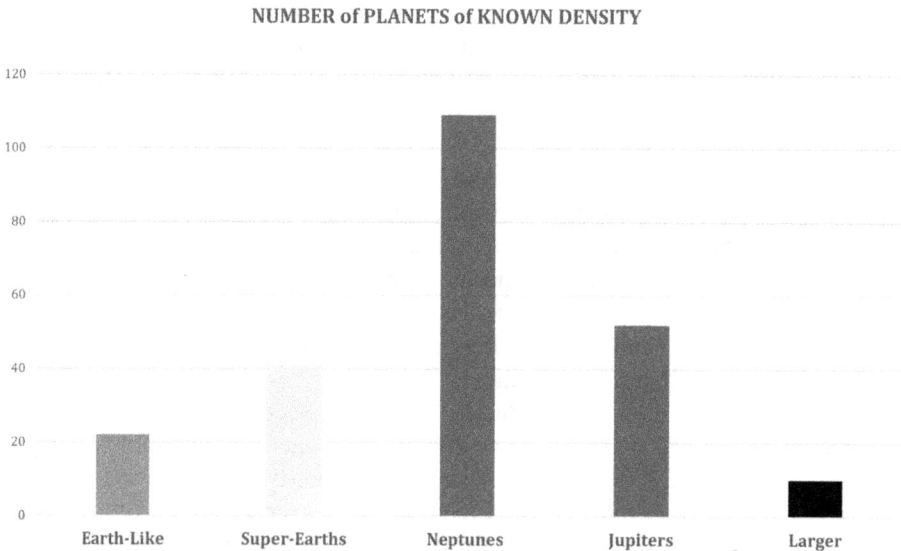

Figure 4.3. Histogram for the data in Figure 4.3 comparing the number of each exoplanet type, as defined by the Kepler mission (Figures 1.7 and 1.8) available online from *NASA's Kepler Mission* (https://exoplanetarchive.ipac.caltech.edu), available in Appendix D or at this book's IOP homepage http://iopscience.iop.org/mono/978-0-7503-6290-0) for which mass and radius are both known.

classification.html should be assigned. This reference introduces mass limits of each exoplanet size category. After this reading, the data for the exoplanets of each separate size category can then be selected out of the complete data set and sorted in order of increasing mass and planets too massive for each size category can be taken out. Doing individual plots of log-density versus radius for each size category, like Figures 4.5–4.8, can then be assigned.

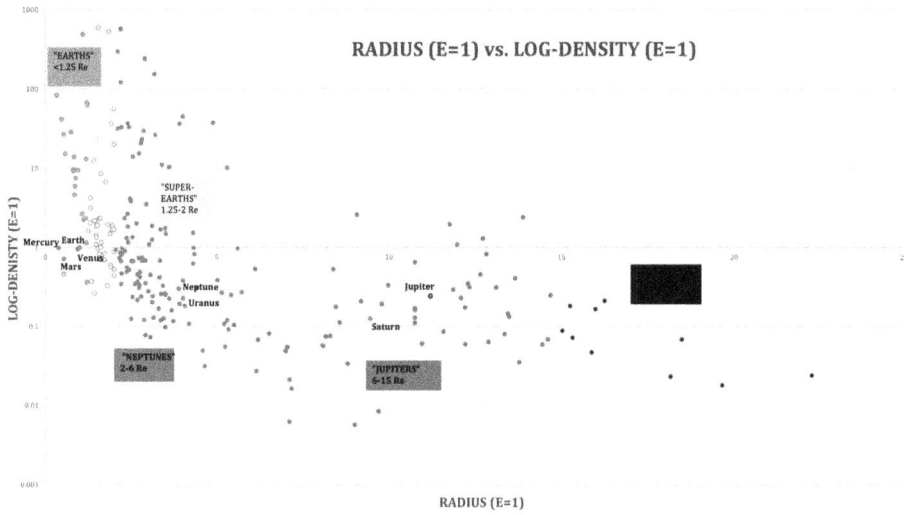

Figure 4.4. A log-plot of the density (Earth density, $E = 1$) as a function of the radii (Earth radius, $E = 1$) of the over 200 Kepler exoplanets (and the planets of our solar system) for which both mass and radius are known.

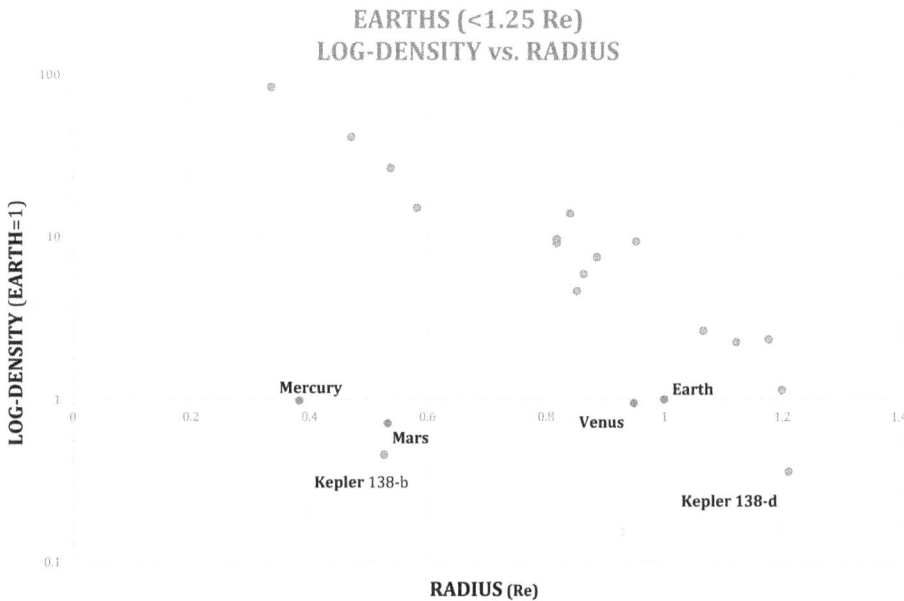

Figure 4.5. Log-plot of the density (Earth density, $E = 1$) as a function of the radii (Earth radius, $E = 1$) for the 17 "*Earths*" (<1.25 * Re, <10 * Me) in the data set and those of our solar system.

SUPER -EARTHS (1.25-2Re)
LOG-DENSITY vs. RADIUS

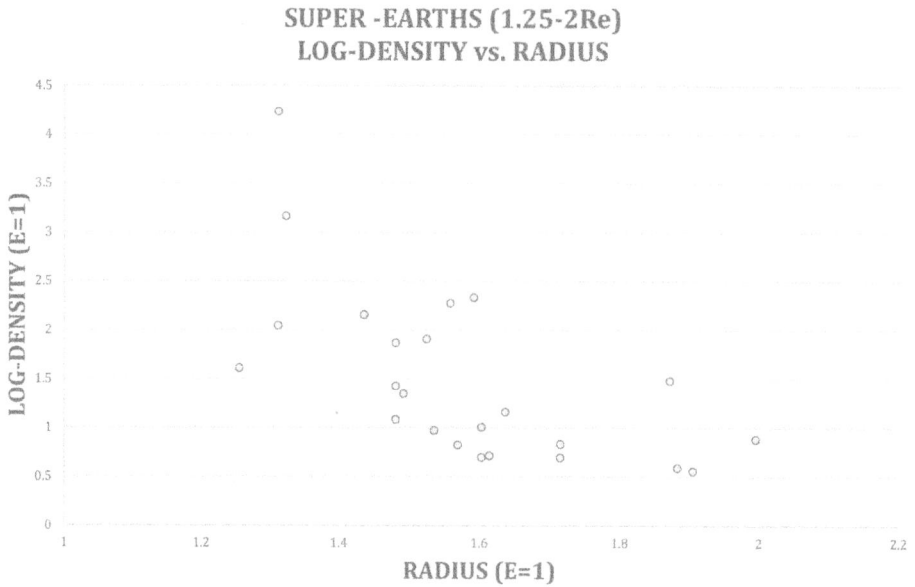

Figure 4.6. Log-plot of the density (Earth density, $E = 1$) as a function of the radii (Earth radius, $E = 1$) for the 29 "*Super-Earths*" (1.25–2 * Re, <10 * Me) in the data set.

NEPTUNES (2-6 Re)
LOG DENSITY vs. RADIUS

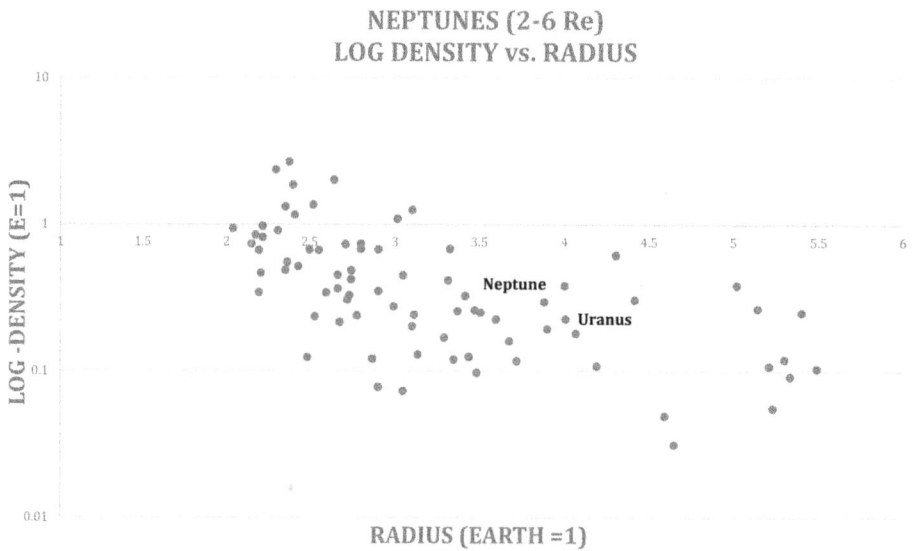

Figure 4.7. Log-plot of the density (Earth density, $E = 1$) as a function of the radii (Earth radius, $E = 1$) for the 78 "*Neptunes*" (2–6 Re, $M < 50$ Me) in the data set.

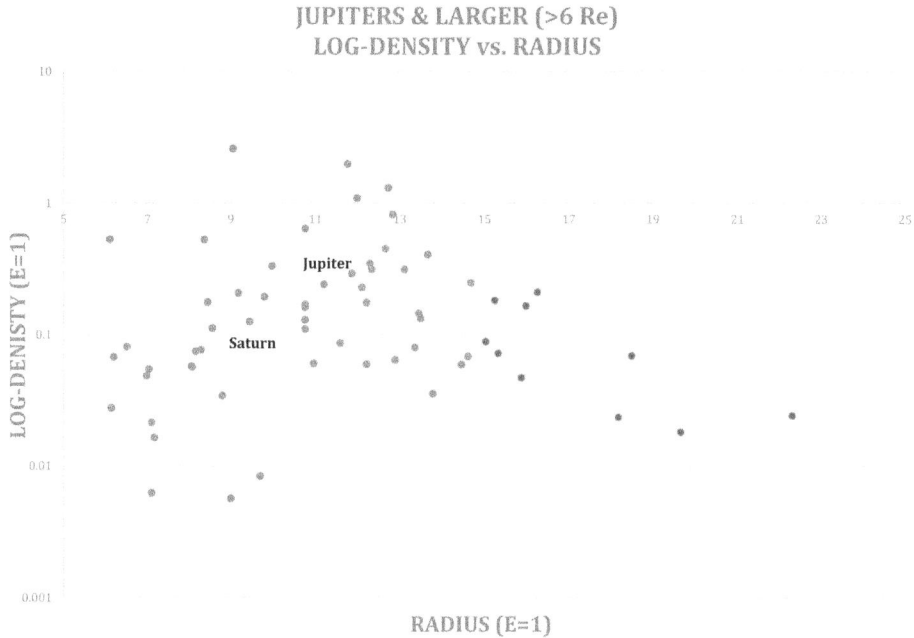

JUPITERS & LARGER (>6 Re)
LOG-DENSITY vs. RADIUS

Figure 4.8. Log-plot of the density (Earth density, $E = 1$) as a function of the radii (Earth radius, $E = 1$) for the 61 *"Jupiter" and "Larger* "(> 6 Re), < 5000 Me) planets in the data set.

4.3 Data and Analysis

Figure 4.4 is a log-plot of densities as a function of radii of all of the over 200 planets from the Kepler mission data set for which mass and radius are both known including the planets of our own solar system for comparison. Based on the known densities of planets in our own solar system (Table 4.1), the general trend seen in the plot of larger planets being less dense is what would be expected. Jupiter and Saturn of our solar system can be seen near the middle of the density range for *"Jupiters"* (6–15 Re). Uranus and Neptune are lower in the density range for *"Neptunes"* (2–6 Re). The plot also shows an overlap of *"Neptunes"* with *"Super-Earths"* (1.25–2 Re). This is where possible *"Mini-Neptunes"* may be found. An attempt to distinguish *"Neptunes"* and *"Super-Earths"* from *"Mini-Neptunes"* based on density is the subject of the study in Chapter 5. The rock–metal terrestrial planets of our solar system also lie in the lower part of the density range for their exoplanet type, *"Earths"* (<1.25 Re).

4.3.1 Earths and Super-Earths

The upper mass limits for *Earths* is considered to be 10 times the mass of Earth, above which a planet is considered a *"Mega-Earth,"* 5 of the planets in the *Earth* radius range (<1.25 Re) were more massive than this and not plotted in Figure 4.5. Most of the *Earths* plotted can be seen to have density higher than those of our solar system suggesting a solid composition. Figure 4.5 also shows 5 *Earths* that are a factor of 10 or more times the density of our Earth, the highest being over 80 times as dense. Two of the planets, however, Kepler 138-d and 138-b, are less dense than Mars. Figure 4.6 is a plot of the planets the *Super-Earth* radius range (1.25–2 Re) with the exception of 12 planets with *Mega-Earth* masses. This plot shows most *Super-Earths* to be of higher density than Earth, and therefore also likely of a solid composition, however, there are 2 less dense than Mars.

4.3.2 Neptunes and Jupiters

Of the 109 exoplanets in the *Neptune* (2–6 Re) radius range, 33 are more than 50 times the mass of Earth, considered the limit for *Neptunes*, leaving 78 plotted in Figure 4.7 along with our solar system's Uranus and Neptune. The plot shows, as would be expected, that most are less dense than Earth. Our solar system's Uranus and Neptune are in the middle of the range. There are a few with a higher density including 17 that are denser than Mars. Most of these denser exoplanets are of radius of about 3 times that of Earth or less, which suggest that they could in fact actually be *Super-Earths* The lower density exoplanets within that radius limit could possibly be classified as *Mini-Neptunes*. These possibilities will be explored in Chapter 5, for which a closer analysis of this portion of the data was done in an attempt to determine the boundary values for both density and radius that separate *Neptunes* and *Super-Earths* from *Mini-Neptunes*.

Of the 62 planets of *Jupiter* (6–15 Re) and *larger* (>15e) size, only 1 is too massive (>5000 Earth masses) for the radius classification. As seen in Figure 4.8 most have similar density to the gas and ice-giants of our solar system. Of those that are denser, only four are of higher density than Earth with one less dense than Earth, but denser than Mars.

4.4 Results

Of the 22 exoplanets with "*Earth*" radii, 5 were of "*Mega-Earth*" masses and 2, Kepler 138-b and 138-d were very small and of lower density than Mars. Although their orbital distance from their star was not reported, this could suggest that they, like many of the exoplanets in the data set for which the orbital distance is known are very close to their star and therefore at a high enough temperature to possibly be "*lava worlds*," composed largely of more molten material. Of exoplanets with "*Super-Earth*" radii 12 of the 41 had "*Mega-Earth*" masses. Of the "*Neptune*" sized exoplanets, 33 of 109 were above the maximum mass limit of the category and 17 were of higher density than Mars, suggesting that they could be "*Super-Earths.*" The above-mentioned study in Chapter 5 is an investigation of possible, mass, radius and density boundaries between the *Neptune*, *Super-Earth*, and *Mini-Neptune* planet types. Of the 62 exoplanets with "*Jupiter*" and "*Larger*" radii only 1 was too massive for the category and only 5 were denser than Mars, 4 of those being denser than Earth. A total of 51 of the 234 exoplanets, almost 22%, were too massive for their size category while, as mentioned above, only a small number of *Earths* and *Jupiters and Larger* exoplanets were found to have densities dissimilar to the others in their category.

4.5 Conclusion

These results are a composite of directed studies, supervised by the author, undertaken by several different undergraduate students during different semesters. The studies completed within a semester when they were taking their introductory astronomy course provided valuable experience for the students by demonstrating several important aspects involved in the process of scientific research, including working with experimental data and the analysis of a larger data set.

Student Projects and Assignments

1 Assign reading of;
   1. *Exoplanet Exploration* -Planets Beyond Our Solar System https://exo-planets.nasa.gov/what-is-an-exoplanet/planet-types/overview/ (or a similar reference) for a student to become familiar with exoplanet detection methods and the different types of exoplanets.
2 Assign reading of;
   1. *Planet Classification*: How to Group Exoplanets https://www.space.com/36935-planet-classification.html (or a similar reference) for a student to become familiar with exoplanet classification by mass.
3 Assign a student to use the data from Appendix D to produce a table and histogram similar to Figures 4.3 and 4.4.
4 Making use of the data from Appendix D, engage a student in a directed study similar to that outlined in this chapter.
5 Assign reading of https://en.wikipedia.org/wiki/Lava_planet to a student who has completed a study similar to the one outlined in this chapter so they can attempt to identify any possible "lava worlds" based on the exoplanet densities.

References and Further Reading

Goldsmith, D. 2018, Exoplanets-Hidden Worlds and the Quest for Extraterrestrial Life (Cambridge, MA: Harvard Univ. Press)

Howell, E. 2017, Planet Classification: How to Group Exoplanets (space.com) https://www.space.com/36935-planet-classification.html

Johnson, J. A. 2016, How Do You Find an Exoplanet (Princeton, NJ: Princeton Univ. Press)

LoPresto, M. C. 2024, *Horizons in World Physics* Vol. 312, ed. A. Reimer (Hauppauge, NY: Nova Publishing)

Tasker, E. 2017, The Planet Factory-Exoplanets and the Search for a Second Earth (New York: Bloomsbury Publishing)

Winn, J. 2023, The Little Book of Exoplanets (Princeton, NJ: Princeton Univ. Press)

Yaqoob, T. 2011, Exoplanets and Alien Solar Systems (Baltimore, MD: New Earth Labs-Education and Outreach)

Chapter 5

Using Densities to Compare Neptunes, Mini-Neptunes and Super-Earths

This study is a comparison of the densities of a subset of extrasolar planets from the Kepler mission data set for which both the radii and masses are known. The comparison is between exoplanets that based on classification by size are considered possible *Super-Earths*, *Mini-Neptunes* or *Neptunes* in an attempt make use of their densities to distinguish which belong in each of these exoplanet categories.

5.1 Introduction

As discussed in previous chapters the *Confirmed Planets* section of the *NASA Exoplanet Archive* includes over 2000 (2311) exoplanets discovered by the NASA Kepler mission, of which over 200 (234) have both a reported radius, determined by the *transit* method (see Figure 1.2) and mass, determined by the *radial velocity* or *Doppler detection* method (see Figure 4.2). This allows a density for these 234 exoplanet to be calculated with equation (4.1), $\rho = 3 \times M/(4 \times \pi r^3)$. The different categories or types of exoplanets are generally classified by size (see Figure 1.7 and Table 1.1), but knowledge of their densities can provide more insight into their possible compositions and may be more useful in attempting to distinguish between *Neptunes*, and two types of exoplanet not actually found in our own solar system: *Super-Earths* and *Mini-Neptunes*.

5.2 Background

The different exoplanet types shown in Figure 1.7 as originally defined by the NASA Kepler mission are grouped by size as detailed in in Table 1.1. This study is concerned with "*Super-Earths*," a "new" type of planet of 1.25–2 Earth radii discovered by the Kepler mission of which there is *not* an example in our solar system; "*Neptunes*" of 2–6 Earth radii and an "overlap" category between larger *Super-Earths*, of possibly up to 4 Earth radii and smaller *Neptunes*, up to only 3 or 4

doi:10.1088/2514-3433/ad7c3ech5

Earth radii and only 20 Earth masses. These are considered "*Mini-Neptunes*," and like *Super-Earths*, there is also no example of a *Mini-Neptunes* in our solar system.

5.3 Procedure and Pedagogy

Whether done as a follow up to the density study in Chapter 4 or as a stand-alone study, this study is not as involved as the previous chapter or the one in Chapter 2, but it is still good practice to limit the tasks assigned to one per meeting.

Also, as with the studies in previous chapters, this one should begin with reading about the topics. Reading of textbook chapter focusing on both the *transit* (Figure 1.2) and *radial velocity* or *Doppler detection* methods (see Figure 4.2) with discussion of both methods to follow. Then, as in the previous studies, reading about the different exoplanets types should occur with *Exoplanet Exploration*—Planets Beyond Our Solar System https://exoplanets.nasa.gov/what-is-an-exoplanet/planet-types/overview/ as a possible source. Specific reading about *Super-Earths* and *Mini-Neptunes* with https://en.wikipedia.org/wiki/Super-Earth and https://en.wikipedia.org/wiki/Mini-Neptune (or similar references) as possible sources. All of the above are suggested as possible **Student Projects and Assignments** below.

5.4 Data and Analysis

As in Chapter 4, the mass and radius spreadsheet data are available in Appendix D (or at this book's IOP homepage, http://iopscience.iop.org/mono/978-0-7503-6290-0) in units of "Jupiters" that must be converted to "Earths", and the density column in "Earth = 1" units must be generated by dividing the mass (in "Earths") column by the cube of the radius (in "Earths") column.

Since an exoplanet could potentially be composed of many different combinations of materials, as mentioned in Chapter 4, it should *not* be assumed that density alone can be used to determine the exact composition of an exoplanet. However, knowledge of the density could be useful for determination of whether exoplanets overlapping between the *Super-Earths* and *Neptune* categories are indeed *Super-Earths* or possibly *Mini-Neptunes*. Being presumably of a more rock–metal composition, *Super-Earths* should have more "Earth-like," higher densities similar to the terrestrial-type planets of our solar system, about the density of Mars, 0.7 × density of Earth, or higher. *Mini-Neptunes* should have lower densities, but perhaps, higher than those of our gas and liquid or icy *Jovian* planets, and *Neptunes* in general, as it has been suggested that their composition may include more water.

The next step is to sort the spreadsheet data by increasing radius and separate out smaller data sets for each of the exoplanet size categories. For this study the data for *Earths* and for *Jupiters* and *larger* exoplanets can be set aside. The data for *Super-Earths* and *Neptunes* can then be sorted by increasing mass. The upper mass limits (https://www.space.com/36935-planet-classification.html), as a reference for classifying exoplanets by mass, for *Super-Earths* (and *Earths*) is considered 10 times the mass of Earth, above which a planet is considered a "*Mega-Earth*." *Mega-Earths* are removed from the data set and also *Neptunes* above their upper mass limit of more than 50 times the mass of Earth. These upper mass limits can be determined by

SUPER -EARTHS
LOG-DENSITY vs. RADIUS

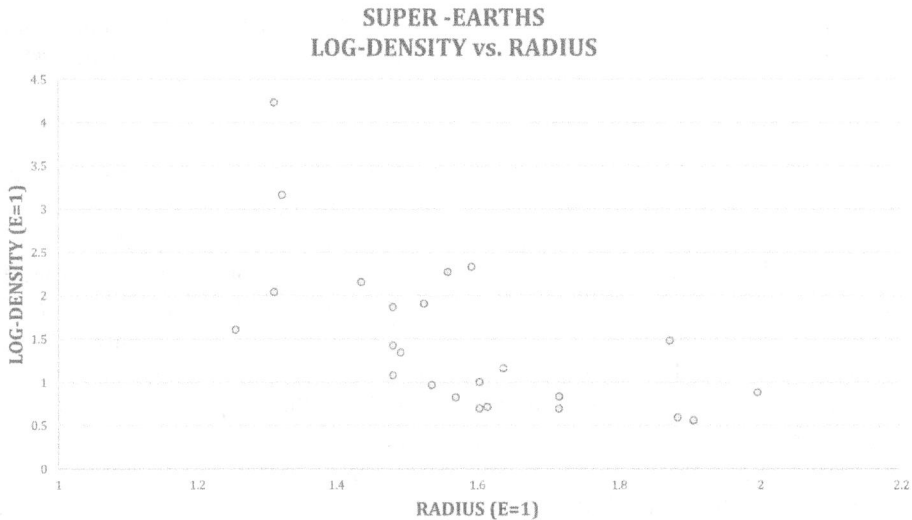

Figure 5.1. Log-plot of the density (Earth density, E = 1) as a function of the radii (Earth radius, E = 1) for the 29 *"Super-Earths"* (1.25–2 × Re, <10 × Me).

reading of the above-mentioned reference, so they do not need to be revealed to the student by the instructor; the student can find them for themselves as part of the process of the study.

Now for the plots. After exoplanets above the upper mass limits are removed, density versus radius should be plotted for the remaining *Super-Earths* and *Neptune*s. Whether the logarithms of the densities or just the densities are plotted is a matter of judgement, log-plots require the extra step of generating a LOG (DENISTY) column, but they can make the differences in the densities more readily visible. Figure 5.1 is a plot of the log-densities versus radii (in units of E = Earth radii) of the 29 *Super-Earths* within the size and mass range. It can be seen that most, as could be expected, are higher density than Earth, and therefore also likely of a similar solid composition, however, two are less dense than Mars.

Figure 5.2 is a plot of the 78 Neptunes within their size and mass range. Most are less dense than Earth. This would be expected based on our solar system's Uranus (0.23 × Earth density) and Neptune (0.3 × Earth density) densities. Some *Neptunes* were of higher density including 17 with a density higher than that of Mars (0.7 × Earth's density). Note that of these higher density exoplanets in the *Neptune* size range are of a radius of about three times that of Earth or less. This suggests that these exoplanets could in fact actually be *Super-Earths* while, the lower density exoplanets within this lower radius limit could be *Mini-Neptunes*. Again, size (and mass) limits for exoplanet types need not necessarily be revealed to the student by instructors; finding them in the give sources could be considered part of their research.

NEPTUNES (2-6 Re)
LOG-DENSITY vs. RADIUS

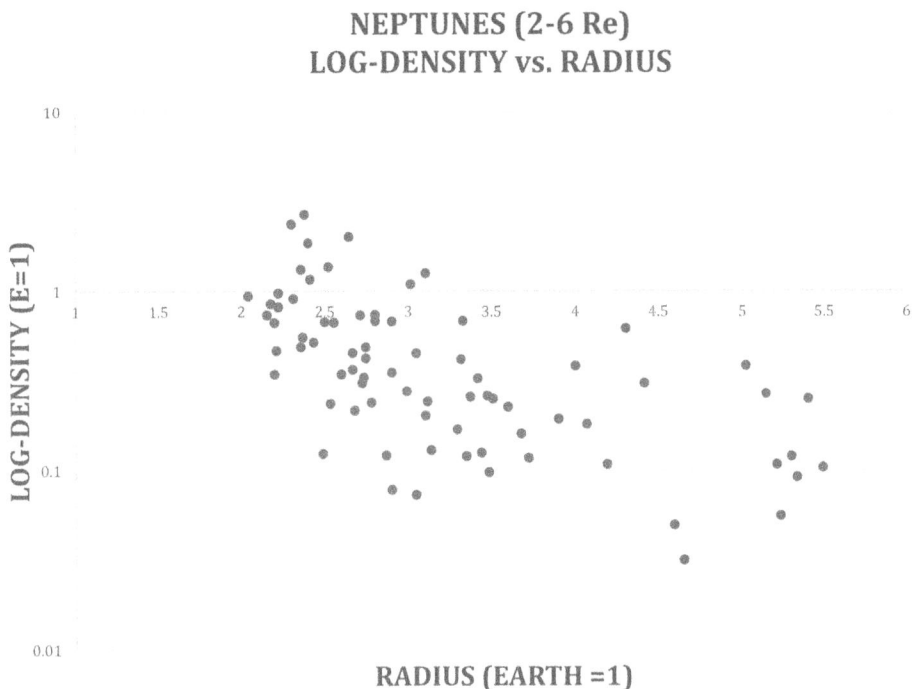

Figure 5.2. Log-plot of the density (Earth density, E = 1) as a function of the radii (Earth radius, E = 1) for the 78 "*Neptunes*" (2–6 Re, *M*<50 Me) in the data set.

5.5 Results

Objects with a radius from 2 to 3 or 4 Earth radii with masses less than 20 Earth masses could be considered *Mini-Neptunes*. The *Super-Earth* size limit can be considered to also be as high as 4 Earth radii. So, there may not actually be a clear dividing line between *Super-Earths* and *Mini-Neptunes* other than that the latter being more gaseous or liquid versions of the former will be less dense. Again, these size and mass limits for exoplanet types need not necessarily be revealed to the student by instructors; finding them in the given sources (or others) could be considered part of the student's research.

Based on the above size and mass limits, a plot, shown in Figure 5.3 for all the possible *Mini-Neptunes* in the data set, can be made. Inspection shows 15 exoplanets of higher density, densities greater than 0.6 × Earth (the least dense being very close to the 0.7 × Earth density of Mars). These might all be *Super-Earths*. The plot also shows 29 exoplanets definitely of lesser density, <0.4 × Earth and also 9 with densities <0. 6 and >0.4 × Earth These could all be *Mini-Neptunes*. The gap at about a density of 0.6 × Earth, could be considered a dividing line between *Super-Earths* and *Mini-Neptunes*.

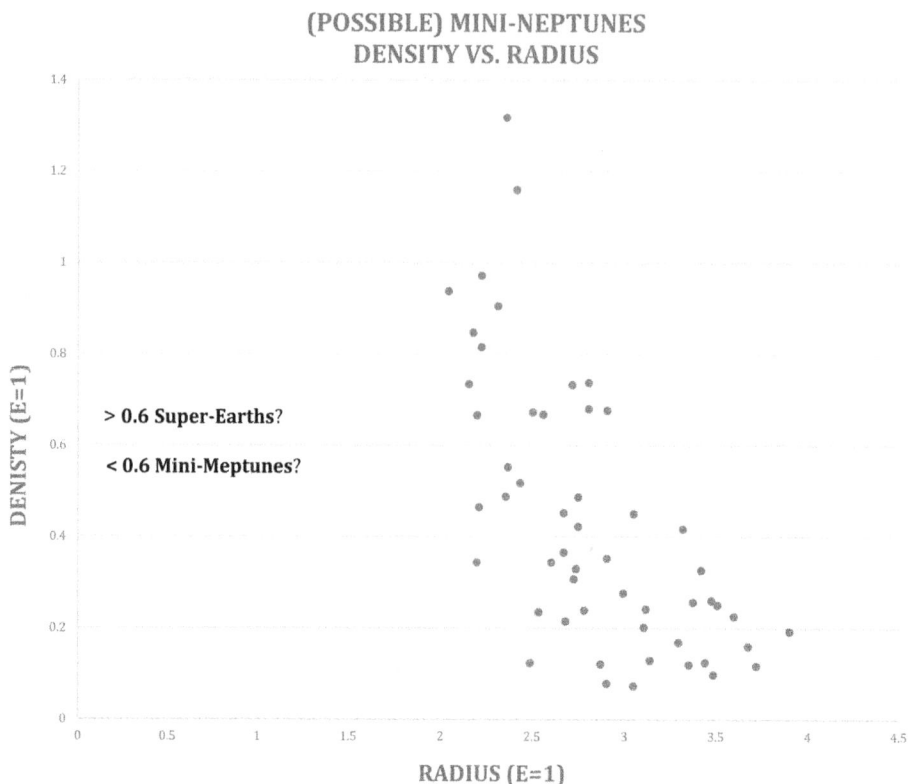

**(POSSIBLE) MINI-NEPTUNES
DENSITY VS. RADIUS**

Figure 5.3. Plot of density vs. radius for the 53 "*Neptunes,*" In the possible "*mini-Neptune*" (2–4 × Re, < 20 × Me) range.

Finally, all of the 86 possible *Super-Earths* and *Mini*-Neptunes from the data set can be plotted together. The plot, Figure 5.4, shows the above-mentioned "gap" from Figure 5.3 at 0.6 × Earth's density that is a possible dividing line between *Super-Earths* and *Mini-Neptunes*. The plot also shows a "radius-cliff" at about 3 Earth radii above which all the exoplanets have lower densities, about 0.4 × Earth density and should be considered *Neptunes*, not *Mini-Neptunes*. This radius-cliff is also evident in Figure 5.3. As mentioned above, it has been suggested that the composition of *Mini-Neptunes* may include more water than the gaseous *Neptunes*, so their density should fall in between that of the more gaseous icy *Neptunes* and the more rocky *Super-Earths*. Kepler 11-b, 128-b and 128-c are anomalies with much lower *Neptune*-like densities, but smaller *Super-Earth* sizes. Although their orbital distances from their star were not reported in the data set, this could suggest that they, like many of the exoplanets in the data set for which the orbital distances were known, are very close to their star and therefore at a high enough temperature to be composed largely of more molten material, known as "lava worlds," as mentioned in the study in Chapter 4.

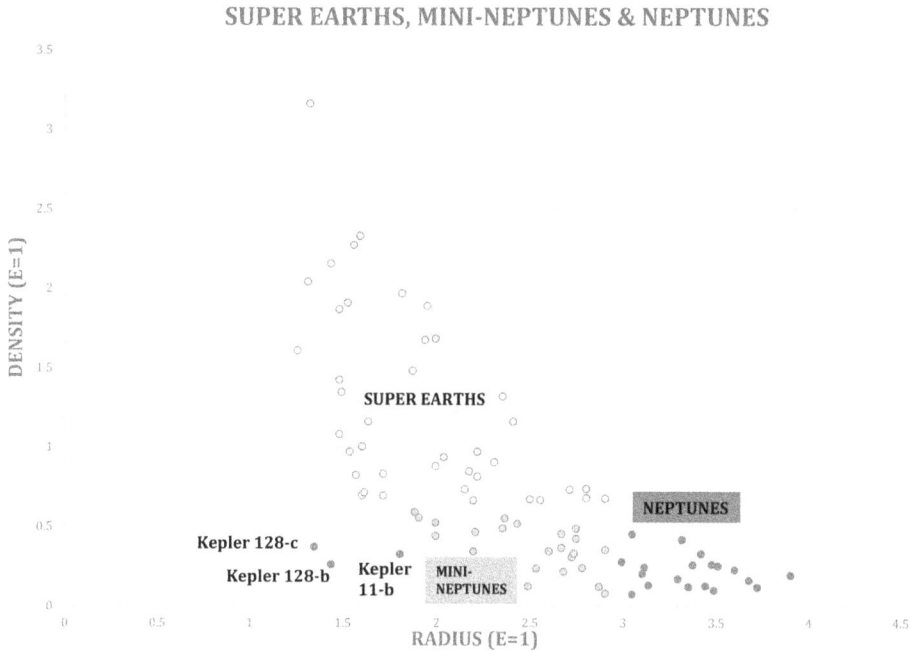

Figure 5.4. Plot of density vs. radius for the 86 planets in the "*Super-Earth*" (1.25–2 × Re) < 10 Me and "*Mini-Neptune*" (2–4 × Re) <20 × Me ranges.

5.6 Conclusion

When comparing potential *Super-Earths* and *Mini-Neptunes*, the plot in Figure 5.3 shows a gap at about 0.6 the density of Earth, suggesting this density as a possible dividing line between the two exoplanet types. Figure 5.4 shows this density gap at 0.6 × Earth's density as well and both plots show the "radius-cliff "at about 3 Earth radii. Above this radius, all the exoplanets are of lower densities and are more likely *Neptune*s. These results verify that density may indeed be a better criterion than size to determine the "gap" between the *Super-Earths* and *Mini-Neptunes.*

A student could certainly be required to finish the studies outlined in these chapters by writing a paper and/or giving a presentation on the results. In general, undertaking this can provide valuable experience for the student by demonstrating and providing practice in some of the skills involved in the analysis of a larger data set, including spreadsheet sorting and calculations as well as the examination of the data and plots, not to mention much more in-depth learning experience than the course alone could provide on one of the most exciting and currently researched topics covered in introductory astronomy, extrasolar planets.

Student Projects and Assignments

1 Assign reading of *Exoplanet Exploration*—Planets Beyond Our Solar System
   https://exoplanets.nasa.gov/what-is-an-exoplanet/planet-types/overview/ (or a

similar reference) for a student to become familiar with exoplanet detection methods and thes different types of exoplanets.

2 Assign reading of https://en.wikipedia.org/wiki/Super-Earth and https://en.wikipedia.org/wiki/Mini-Neptune (or a similar references) for a student to become familiar with *Super-Earths* and *Mini-Neptunes* and especially their differences.

3 Assign reading of *Planet Classification*: How to Group Exoplanets https://www.space.com/36935-planet-classification.html (or a similar reference) for a student to become familiar with exoplanet classification by mass.

4 Making use of the data from **Appendix D**, engage a student in a directed study similar to that outlined in this chapter.

5 **Appendix E** provides a faster, more guided, activity for students to use the data discovered by the Kepler mission to compare the densities of exoplanets to those of similar size-classifications in our own solar system and make estimates of the size and density boundaries between *Super-Earths*, *Mini-Neptunes*, and *Neptunes*.

References

https://en.wikipedia.org/wiki/Lava_planet

https://en.wikipedia.org/wiki/Mini-Neptune

https://en.wikipedia.org/wiki/Super-Earth

Exoplanet Exploration—Planets Beyond Our Solar System https://exoplanets.nasa.gov/what-is-an-exoplanet/planet-types/overview/

Goldsmith, D. 2018, Exoplanets-Hidden Worlds and the Quest for Extraterrestrial Life (Cambridge, MA: Harvard Univ. Press)

Howell, E. 2017, Planet Classification: How to Group Exoplanets, https://www.space.com/36935-planet-classification.html

Johnson, J. A. 2016, How Do You Find an Exoplanet (Princeton, NJ: Princeton Univ. Press)

LoPresto, M. C. 2024, *Horizons in World Physics* Vol. 312, ed. A. Reimer (Hauppauge, NY: Nova Publishing)

Luque, R., & Palle, E. 2022, Sci, 377, 1211

Tasker, E. 2017, The Planet Factory-Exoplanets and the Search for a Second Earth (New York: Bloomsbury Publishing)

Yaqoob, T. 2011, Exoplanets and Alien Solar Systems (Baltimore, MD: New Earth Labs-Education and Outreach)

Chapter 6

Beyond the Kepler Mission

The discoveries of the Kepler mission are just the beginning of the search for habitable exoplanets. The *Transiting Exoplanet Survey Satellite* (*TESS*) has already confirmed discoveries of exoplanets and is expected to find many smaller exoplanets within the habitable zones of their stars. Some of these will become prime targets for the *James Webb Space Telescope (JWST)* that will analyze the spectra of the atomspheres of these exoplanets for chemical signatures that could indicate habitablity and possibly even life.

6.1 Kepler's Discoveries

After nearly a decade of operation the Kepler space telescope https://science.nasa.gov/mission/kepler was officially retired on 2018 October 30. This was also the 338th anniversary of the death of its namesake, the astronomer Johannes Kepler (Figure 6.1) in 1630. Through use of the *transit-method*, the Kepler mission discovered over 2000 exoplanets, about half of which were much smaller, *Earths* and *Super-Earths*, a planet type not previously known of which there is no example in our own solar system. Among these were about a dozen that were considered potentially habitable https://phl.upr.edu/projects/habitable-exoplanets-catalog.

6.2 Habitability

Whether or not a planet lies in its star's habitable zone is only the very first step in determining whether or not a planet is really habitable. This is why Earth-sized planets in habitable zones are only referred to as *potentially* habitable.

For water to exist as a liquid, not only do temperatures have to be in the correct range, but there also must be enough pressure to stop the liquid water from evaporating into gas. On Earth this pressure is provided by the atmosphere. Put simply, the atmospheric pressure at any point on or above the surface of a planet is caused by the weight of the air overhead. So in order to provide the necessary pressure for water to be liquid a planet must be massive enough to have enough

doi:10.1088/2514-3433/ad7c3ech6   6-1

Figure 6.1. Johannes Kepler (1571–1630). German Astronomer who discovered the Three Laws of Planetary Motion.

gravitational pull to retain a substantial atmosphere. In our solar system, Earth and Venus are massive enough, while smaller objects like Mercury and Earth's moon and most of the moons of the outer planets are not. The presence of an atmosphere will also protect potential life on a planet from harmful radiation from space as well as many impacts.

When a planet first forms from collisions of smaller objects in orbit of a star, the heat from the collisions causes the planet to be in a molten state. The molten material then cools, giving off gases. This volcanic outgassing, as it is called, is how a planet's intial atmopshere forms. The cooling occurs from the outside to the inside of the planet and as long as there is still a substanital molten core, there will be geologic activity which is also very important for habitability.

Volcanic eruptions replenish an atmosphere with the same gases that were released when outgassing first began. Motions of a planet's crustal plates, tectonics, regulate the carbon cycle, carbon dioxide in the atompshere being absorbed in water, condensing in carbonaceous rocks and then being rereleased back into the atmopshere. The amount of carbon dioxide in an atmosphere will affect the amount of greenhouse effect (see Chapter 3) that will occur in a planet's atmosphere. The carbon cycle on Earth keeps the amount carbon dioxide in Earth's atmosphere much less than on Venus, where there are volcanic eruptions and a very thick atmosphere, but no tectonics. This results in temperatures much too high for water to be liquid. Much smaller and less massive Mars has a much thinner atmosphere and is very cold so its water is frozen. Mars being much smaller than Earth, has already cooled off and is no longer geologically active, but inactive volcanoes and tectonic rift valleys provide evidence that it once was. This means that its atmosphere was being replenished and was likely thicker, exerting more pressure. More pressure and greehouse effect heating, allowed for liquid water, as is suggested by the presences of dried up river beds.

Electric charges in a molten core can also provide a planet, if it rotates fast enough, with a magnetic field. A magnetic field will also provide protection from harmful radiation from space and can help prevent an atmosphere from being blown away to space by stellar (or solar) winds, charged particles emitted from processes that occur within a star. Mars rotates at about the same rate as Earth, so in the past, when it was geologically active, it had a magnetic field and was much more habitable than it is today. Friction with its thick atmosphere causes Venus to rotate very slowly and therefore, despite still having a molten core, Venus does not have a magnetic field. This and its very high temperatures causes Venus not to be habitable.

All planets will eventually cool off, larger ones more slowly than smaller ones. No longer having a molten core and thus no longer being geologically active will render a planet no longer habitable. Along with distance from its star, mass and size, and the rotation rate of a planet, this makes the age of a planet another very important factor in habitablity.

6.3 Ongoing and Future Missions

Several missions picking up where Kepler left off, continuing the search for potentially habitable exoplanets are the *Transiting Exoplanet Survey Satellite* (*TESS*) https://exoplanets.nasa.gov/tess/ launched in 2018 from which there is already data available in the *NASA Exoplanet Archive* (https://exoplanetarchive. ipac.caltech.edu), the *James Webb Space Telescope (JWST)* finally launched near the end of 2021 after many delays, and the soon to be launched *Nancy Grace Roman Space Telescope* https://roman.gsfc.nasa.gov.

For its first two years of operation, TESS (Figure 6.2) searched for tranisting exoplanets orbiting brighter stars near Earth, stars much closer than those investiaged by the Kepler mission. TESS was also desgined to survey 85% of the sky using wide-field cameras. At the time of this writing the discovery of over 400 exoplanets discovered by TESS have already been confimed and it is expected to find large numbers of small planets in the habitable zones of their stars for which size and mass and therefore density can be studied. TESS is also expected to discover exoplanets that will be targets for further investigation by the *James Webb Space Telescope* (JWST) https://science.nasa.gov/mission/webb/. The portion of JWST 's mission that is focused on exoplanets is spectoroscopic examination of exoplanet atmospheres at infrared wavelengths and looking for spectroscopic signatures of chemicals that could be conducive to the presence of life, such as water, carbon dioxide, nitrogen and some that may even suggest the presense of life such as oxygen and methane (Figure 6.3).

Figure 6.2. Artist's conception of the *Transiting Exoplanet Survey Satellite* (*TESS*) (Courtesy of NASA images, http://www.nasa.gov).

Figure 6.3. Artist's conception of the *James Web Space Telescope (JWST)* (Courtesy of NASA images, http://www.nasa.gov).

The *Nancy Grace Roman Space Telescope*, scheduled for launch later this decade, will spend a 5–10 year mission searching for exoplanets using gravitaitonal microlensing, https://exoplanets.nasa.gov/resources/2168/gravitational-microlensing/. It is expected to find at least as many exoplanets or more than the Kepler mission, including some with masses as low as between those of Mars and our Moon. The mission "*Roman*," for short, as it will be called, as did its predecessor, the Kepler mission, will also focus on questions about habitablity and the possible presence of life.

6.4 Conclusion

The discoveries of the Kepler mission have greatly increased our knowledge about exoplanets and their possible habitability, but as indicated it is has really only been a beginning. The ongoing and future missions discussed and certainly others in coming years and decades will continue the search, the next steps being identifying exoplanets with atmospheres that could support life and possibly even the presense of life itself. Which would finally be an anwer to perhaps the single most important question ever asked by the human species, "are we alone?"

Student Projects and Assignments

1 Assign a student to research the TESS (https://exoplanets.nasa.gov/tess/, JWST https://science.nasa.gov/mission/webb/ or ROMAN https://roman.gsfc.nasa.gov mission and report on the their scientific goals and/or results related to exoplanets.
2 Assign a student to make use of the exoplanet data for from the *TESS*-mission (available in Appendix F or at this book's IOP homepage, http://iopscience.iop.org/mono/978-0-7503-6290-0), the data is from the *NASA Exoplanet Archive* https://exoplanetarchive.ipac.caltech.edu, to do a study similar to one or more of those that were outlined with the Kepler mission data in previous chapters.

References and Further Reading

Crockett, C. 2016, SciN, 187, 32

Redd, N. T. 2016, Ast, 44, 25

Wall, M. 2016, Alien Atmospheres: The Search for Signs of Life (Space.com) https://www.space.com/31519-alien-life-hunt-biosignatures-exoplanet-atmospheres.html

Wenz, J. 2017, Ast, 45, 44

Winn, J. 2023, The Little Book of Exoplanets (Princeton, NJ: Princeton Univ. Press)

Student Exoplanet Projects Using Data from the Kepler Mission

Michael C LoPresto

Appendix A

Kepler Mission Data Spreadsheet

This appendix is also available as a spreadsheet from [doi link 10.1088/978-0-7503-6290-0]

Host	Letter	Number	p (days)	d (AU)	M (Earths)	R (Earths)	Temperature (K)	M_s (Suns)	R_s (Suns)
Kepler-10	b	2	0.84	0.0172	4.611	1.4784	5627	0.91	1.06
Kepler-101	b	2	3.49	0.0474	50.88	5.712	5667	1.17	1.56
Kepler-101	c	2	6.03	0.0684	3.18	1.2544	5667	1.17	1.56
Kepler-104	b	3	11.43	0.094	0	3.1024	5711	0.81	1.35
Kepler-104	c	3	23.67	0.153	0	3.1248	5711	0.81	1.35
Kepler-104	d	3	51.76	0.257	0	3.5728	5711	0.81	1.35
Kepler-105	b	2	5.41	0.066	0	4.8048	5827	0.96	0.89
Kepler-107	b	4	3.18	0.044	0	1.5568	5851		1.41
Kepler-107	c	4	4.9	0.059	0	1.8032	5851		1.41
Kepler-107	d	4	7.96	0.082	0	1.064	5851		1.41
Kepler-107	e	4	14.75	0.123	0	3.4496	5851		1.41
Kepler-108	b	2	49.18	0.292	0	8.6464	5854	0.87	2.19
Kepler-108	c	2	190.32	0.721	0	8.176	5854	0.87	2.19
Kepler-11	b	6	10.3	0.091	1.908	1.8032	5663	0.96	1.06
Kepler-11	c	6	13.02	0.107	2.862	2.8672	5663	0.96	1.06
Kepler-11	d	6	22.68	0.155	7.314	3.1136	5663	0.96	1.06
Kepler-11	e	6	32	0.195	7.95	4.1888	5663	0.96	1.06
Kepler-11	f	6	46.69	0.25	1.908	2.4864	5663	0.96	1.06
Kepler-11	g	6	118.38	0.466	25.122	3.3264	5663	0.96	1.06
Kepler-110	b	2	12.69	0.107	0	1.8256	5960		1.15
Kepler-110	c	2	31.72	0.198	0	2.2064	5960		1.15
Kepler-111	b	2	3.34	0.046	0	1.5792	5952		1.16
Kepler-111	c	2	224.78	0.761	0	7.2912	5952		1.16
Kepler-112	b	2	8.41	0.076	0	2.3632	5544		0.84
Kepler-112	c	2	28.57	0.172	0	2.3968	5544		0.84
Kepler-114	b	3	5.19	0.053	0	1.2544	4605	0.56	0.67
Kepler-115	b	2	2.4	0.036	0	1.0864	5979	1	1.21
Kepler-115	c	2	8.99	0.087	0	2.5984	5979	1	1.21

(Continued)

(*Continued*)

Host	Letter	Number	p (days)	d (AU)	M (Earths)	R (Earths)	Temperature (K)	M_s (Suns)	R_s (Suns)
Kepler-116	b	2	5.97	0.069	0	3.416	6142	1.16	1.45
Kepler-116	c	2	13.07	0.116	0	2.296	6142	1.16	1.45
Kepler-117	b	2	18.8	0.1445	29.892	8.0528	6150	1.13	1.61
Kepler-117	c	2	50.79	0.2804	585.12	12.3312	6150	1.13	1.61
Kepler-118	b	2	7.52	0.073	0	1.96	5274	0.86	1.09
Kepler-118	c	2	20.17	0.141	0	7.672	5274	0.86	1.09
Kepler-119	b	2	2.42	0.035	0	3.5952	5595		0.84
Kepler-119	c	2	4.13	0.049	0	0.9184	5595		0.84
Kepler-12	b	1	4.44	0.0553	137.376	19.6448	5947	1.17	1.48
Kepler-120	b	2	6.31	0.055	0	2.1504	4096		0.53
Kepler-120	c	2	12.79	0.088	0	1.5232	4096		0.53
Kepler-121	b	2	3.18	0.039	0	2.3408	5311		0.7
Kepler-121	c	2	41.01	0.216	0	2.2736	5311		0.7
Kepler-122	b	5	5.77	0.064	0	2.3408	6050	0.99	1.22
Kepler-122	c	5	12.47	0.108	0	5.8688	6050	0.99	1.22
Kepler-122	d	5	21.59	0.155	0	2.1952	6050	0.99	1.22
Kepler-122	e	5	37.99	0.227	0	2.5984	6050	0.99	1.22
Kepler-123	b	2	17.23	0.135	0	2.9344	6089	1.03	1.26
Kepler-123	c	2	26.7	0.181	0	1.4784	6089	1.03	1.26
Kepler-124	b	3	3.41	0.039	0	0.728	4984		0.64
Kepler-124	c	3	13.82	0.1	0	1.7472	4984		0.64
Kepler-124	d	3	30.95	0.17	0	1.1088	4984		0.64
Kepler-125	b	2	4.16	0.041	0	2.3632	3810	0.55	0.51
Kepler-125	c	2	5.77	0.051	0	0.7392	3810	0.55	0.51
Kepler-126	b	3	10.5	0.099	0	1.5232	6239		1.36
Kepler-126	c	3	21.87	0.162	0	1.5792	6239		1.36
Kepler-126	d	3	100.28	0.448	0	2.4976	6239		1.36
Kepler-127	b	3	14.44	0.125	0	1.4	6106		1.36
Kepler-127	c	3	29.39	0.2	0	2.6544	6106		1.36
Kepler-127	d	3	48.63	0.28	0	2.6432	6106		1.36
Kepler-129	b	2	15.79	0.131	0	2.3632	5770	1.18	1.64
Kepler-129	c	2	82.2	0.393	0	2.5424	5770	1.18	1.64
Kepler-130	b	3	8.46	0.079	0	1.0192	5884	1	1.13
Kepler-130	c	3	27.51	0.178	0	2.9008	5884	1	1.13
Kepler-130	d	3	87.52	0.377	0	1.6352	5884	1	1.13
Kepler-132	b	4	6.18	0.067	0	1.2096	6003		1.18
Kepler-132	c	4	6.41	0.068	0	1.2768	6003		1.18
Kepler-132	d	4	18.01	0.136	0	1.5456	6003		1.18
Kepler-133	b	2	8.13	0.083	0	1.7584	5736		1.43
Kepler-133	c	2	31.52	0.204	0	2.8336	5736		1.43
Kepler-134	b	2	5.32	0.06	0	1.9936	5983		1.18
Kepler-134	c	2	10.11	0.092	0	1.2544	5983		1.18
Kepler-135	b	2	6	0.067	0	1.8032	6090		1.27
Kepler-135	c	2	11.45	0.103	0	1.1536	6090		1.27
Kepler-136	b	2	11.58	0.106	0	2.0496	6165	1.2	1.35
Kepler-136	c	2	16.4	0.133	0	1.9936	6165	1.2	1.35

Kepler-137	b	2	8.44	0.077	0	1.4672	5187		0.8
Kepler-137	c	2	18.74	0.13	0	1.8816	5187		0.8
Kepler-139	b	2	15.77	0.127	0	2.9344	5594	1.08	1.3
Kepler-139	c	2	157.07	0.586	0	3.3824	5594	1.08	1.3
Kepler-140	b	2	3.25	0.045	0	1.6128	6077		1.29
Kepler-140	c	2	91.35	0.414	0	1.8032	6077		1.29
Kepler-141	b	2	3.11	0.039	0	0.6944	4910	1	0.79
Kepler-141	c	2	7.01	0.067	0	1.4112	4910	1	0.79
Kepler-142	b	3	2.02	0.032	0	1.9936	5790	0.99	1.27
Kepler-142	c	3	4.76	0.057	0	2.856	5790	0.99	1.27
Kepler-142	d	3	41.81	0.242	0	2.1616	5790	0.99	1.27
Kepler-143	b	2	16.01	0.127	0	2.408	5848		1.36
Kepler-143	c	2	27.08	0.181	0	3.3712	5848		1.36
Kepler-144	b	2	5.89	0.066	0	1.3328	6075	1.03	1.24
Kepler-144	c	2	10.1	0.094	0	1.344	6075	1.03	1.24
Kepler-146	b	2	31.16	0.2	0	3.7072	5948		1.21
Kepler-146	c	2	76.73	0.364	0	3.1248	5948		1.21
Kepler-147	b	2	12.61	0.113	0	1.5232	6012	1.01	1.47
Kepler-147	c	2	33.42	0.216	0	2.4304	6012	1.01	1.47
Kepler-148	b	3	1.73	0.028	0	1.8032	5272		0.85
Kepler-148	c	3	4.18	0.05	0	3.5952	5272		0.85
Kepler-149	b	3	29.2	0.184	0	4.2112	5381		0.95
Kepler-149	c	3	55.33	0.281	0	1.6128	5381		0.95
Kepler-149	d	3	160.02	0.571	0	3.9536	5381		0.95
Kepler-15	b	1	4.94	0.05714	209.88	10.752	5515	1.02	0.99
Kepler-150	b	4	3.43	0.044	0	1.2544	5560		0.94
Kepler-150	c	4	7.38	0.073	0	3.6848	5560		0.94
Kepler-150	d	4	12.56	0.104	0	2.7888	5560		0.94
Kepler-150	e	4	30.83	0.189	0	3.1136	5560		0.94
Kepler-151	b	2	15.23	0.116	0	3.0576	5460	0.83	0.83
Kepler-151	c	2	24.67	0.16	0	2.0832	5460	0.83	0.83
Kepler-152	b	2	18.21	0.124	0	2.7888	5088		0.72
Kepler-152	c	2	88.26	0.356	0	2.3856	5088		0.72
Kepler-153	b	2	18.87	0.129	0	2.9232	5404		0.89
Kepler-153	c	2	46.9	0.237	0	2.5312	5404		0.89
Kepler-154	b	5	33.04	0.198	0	2.2624	5690	0.89	1
Kepler-154	c	5	62.3	0.303	0	2.9456	5690	0.89	1
Kepler-155	b	2	5.93	0.056	0	2.0832	4508	0.58	0.62
Kepler-155	c	2	52.66	0.242	0	2.24	4508	0.58	0.62
Kepler-156	b	2	4.97	0.054	0	2.296	5094		0.81
Kepler-156	c	2	15.91	0.117	0	2.5424	5094		0.81
Kepler-157	b	3	1.73	0.028	0	1.3216	5774		1.04
Kepler-157	c	3	13.54	0.11	0	2.24	5774		1.04
Kepler-158	b	2	16.71	0.111	0	2.1168	4623		0.62
Kepler-158	c	2	28.55	0.158	0	1.904	4623		0.62
Kepler-159	b	2	10.14	0.082	0	2.3744	4625		0.66
Kepler-159	c	2	43.6	0.218	0	3.4048	4625		0.66
Kepler-16	b	1	228.78	0.7048	105.894	8.4448	4450	0.69	0.65

(Continued)

(*Continued*)

Host	Letter	Number	p (days)	d (AU)	M (Earths)	R (Earths)	Temperature (K)	M_s (Suns)	R_s (Suns)
Kepler-160	b	2	4.31	0.05	0	1.5344	5857		0.88
Kepler-160	c	2	13.7	0.109	0	3.6064	5857		0.88
Kepler-165	b	2	8.18	0.072	0	2.3184	5211		0.77
Kepler-165	c	2	15.31	0.11	0	2.2288	5211		0.77
Kepler-166	b	3	7.65	0.072	0	2.2736	5413		0.74
Kepler-166	c	3	34.26	0.195	0	2.3744	5413		0.74
Kepler-167	b	4	4.39	0.0483	0	1.6128	4890	0.77	0.73
Kepler-167	c	4	7.41	0.0684	0	1.5456	4890	0.77	0.73
Kepler-167	d	4	21.8	0.1405	0	1.1984	4890	0.77	0.73
Kepler-167	e	4	1071.23	1.89	0	10.1472	4890	0.77	0.73
Kepler-168	b	2	4.43	0.056	0	1.456	6282		1.11
Kepler-168	c	2	13.19	0.116	0	2.688	6282		1.11
Kepler-169	b	5	3.25	0.04	0	1.1312	4997	0.86	0.76
Kepler-169	c	5	6.2	0.062	0	1.2096	4997	0.86	0.76
Kepler-169	d	5	8.35	0.075	0	1.2544	4997	0.86	0.76
Kepler-169	e	5	13.77	0.105	0	2.1952	4997	0.86	0.76
Kepler-169	f	5	87.09	0.359	0	2.576	4997	0.86	0.76
Kepler-17	b	1	1.49	0.02591	779.1	14.672	5781	1.16	1.05
Kepler-170	b	2	7.93	0.08	0	3.192	5679		1.03
Kepler-170	c	2	16.67	0.131	0	2.856	5679		1.03
Kepler-171	b	3	4.17	0.05	0	2.3408	5642		0.84
Kepler-171	c	3	11.46	0.098	0	2.5536	5642		0.84
Kepler-171	d	3	39.6	0.223	0	1.8928	5642		0.84
Kepler-172	b	4	2.94	0.04	0	2.352	5526	0.86	1.08
Kepler-172	c	4	6.39	0.068	0	2.856	5526	0.86	1.08
Kepler-172	d	4	14.63	0.118	0	2.2512	5526	0.86	1.08
Kepler-172	e	4	35.12	0.211	0	2.7552	5526	0.86	1.08
Kepler-173	b	2	4.26	0.048	0	1.288	6031	0.78	0.95
Kepler-173	c	2	8.01	0.074	0	2.4304	6031	0.78	0.95
Kepler-174	b	3	13.98	0.1	0	1.96	4880		0.62
Kepler-174	c	3	44	0.214	0	1.4896	4880		0.62
Kepler-174	d	3	247.35	0.677	0	2.184	4880		0.62
Kepler-175	b	2	11.9	0.105	0	2.5536	6064	1.04	1.01
Kepler-175	c	2	34.04	0.213	0	3.1024	6064	1.04	1.01
Kepler-176	b	4	5.43	0.058	0	1.4336	5232		0.89
Kepler-176	c	4	12.76	0.102	0	2.5984	5232		0.89
Kepler-176	d	4	25.75	0.163	0	2.5088	5232		0.89
Kepler-178	b	3	9.58	0.085	0	2.9008	5676		1.07
Kepler-178	c	3	20.55	0.142	0	2.8784	5676		1.07
Kepler-178	d	3	96.68	0.397	0	3.9424	5676		1.07
Kepler-179	b	2	2.74	0.036	0	1.6352	5302		0.76
Kepler-179	c	2	6.4	0.064	0	1.9936	5302		0.76
Kepler-18	b	3	3.5	0.0447	6.996	1.9936	5345	0.97	1.11
Kepler-18	c	3	7.64	0.0752	17.172	5.488	5345	0.97	1.11
Kepler-18	d	3	14.86	0.1172	16.536	6.9776	5345	0.97	1.11
Kepler-180	b	2	13.82	0.109	0	1.5008	5731	0.84	1.06

Kepler-180	c	2	41.89	0.229	0	3.0128	5731	0.84	1.06
Kepler-181	b	2	3.14	0.04	0	1.2656	5333		0.75
Kepler-181	c	2	4.3	0.049	0	1.9936	5333		0.75
Kepler-182	b	2	9.83	0.096	0	2.576	6250	1.14	1.15
Kepler-182	c	2	20.68	0.157	0	3.4272	6250	1.14	1.15
Kepler-183	b	2	5.69	0.064	0	2.0608	5888		0.96
Kepler-183	c	2	11.64	0.103	0	2.2736	5888		0.96
Kepler-184	b	3	10.69	0.092	0	2.3632	5788		0.87
Kepler-184	c	3	20.3	0.141	0	1.9712	5788		0.87
Kepler-184	d	3	29.02	0.179	0	2.4864	5788		0.87
Kepler-185	b	2	1.63	0.026	0	1.1648	5208	0.79	0.81
Kepler-185	c	2	20.73	0.139	0	2.016	5208	0.79	0.81
Kepler-186	b	5	3.89	0.0343	0	1.064	3755	0.54	0.52
Kepler-186	c	5	7.27	0.0451	0	1.2544	3755	0.54	0.52
Kepler-186	d	5	13.34	0.0781	0	1.4	3755	0.54	0.52
Kepler-186	e	5	22.41	0.11	0	1.2656	3755	0.54	0.52
Kepler-186	f	5	129.94	0.432	0	1.1648	3755	0.54	0.52
Kepler-187	b	2	4.94	0.059	0	1.4112	6105	0.85	1.29
Kepler-187	c	2	10.64	0.099	0	2.6656	6105	0.85	1.29
Kepler-188	b	2	2.06	0.032	0	1.68	6021		1.14
Kepler-188	c	2	6	0.066	0	3.192	6021		1.14
Kepler-189	b	2	10.4	0.088	0	1.2096	5235	0.79	0.75
Kepler-189	c	2	20.13	0.137	0	2.3744	5235	0.79	0.75
Kepler-190	b	2	2.02	0.03	0	1.5568	5106	0.84	0.8
Kepler-190	c	2	3.76	0.045	0	1.456	5106	0.84	0.8
Kepler-191	b	3	9.94	0.087	0	1.344	5282	0.85	0.79
Kepler-191	c	3	17.74	0.128	0	1.8592	5282	0.85	0.79
Kepler-192	b	3	9.93	0.09	0	2.7328	5479		1.01
Kepler-192	c	3	21.22	0.15	0	2.7888	5479		1.01
Kepler-193	b	2	11.39	0.106	0	2.3856	6335		1.15
Kepler-193	c	2	50.7	0.286	0	2.744	6335		1.15
Kepler-194	b	3	2.09	0.032	0	1.512	6089		1.02
Kepler-194	c	3	17.31	0.131	0	2.5872	6089		1.02
Kepler-194	d	3	52.81	0.275	0	2.3968	6089		1.02
Kepler-195	b	2	8.31	0.077	0	2.0272	5329		0.78
Kepler-195	c	2	34.1	0.197	0	1.5456	5329		0.78
Kepler-196	b	2	20.74	0.138	0	1.904	5128		0.78
Kepler-196	c	2	47.43	0.24	0	2.24	5128		0.78
Kepler-197	b	4	5.6	0.06	0	1.0192	6004		1.12
Kepler-197	c	4	10.35	0.09	0	1.232	6004		1.12
Kepler-197	d	4	15.68	0.119	0	1.2208	6004		1.12
Kepler-197	e	4	25.21	0.164	0	0.9072	6004		1.12
Kepler-198	b	3	17.79	0.131	0	2.8224	5574	0.93	0.94
Kepler-198	c	3	49.57	0.259	0	2.464	5574	0.93	0.94
Kepler-199	b	2	23.64	0.158	0	3.1024	5644		0.97
Kepler-199	c	2	67.09	0.316	0	3.248	5644		0.97
Kepler-20	b	5	3.7	0.04537	8.586	1.904	5455	0.91	0.94
Kepler-20	c	5	10.85	0.093	16.218	3.0688	5455	0.91	0.94
Kepler-20	d	5	77.61	0.3453	20.034	2.744	5455	0.91	0.94

(Continued)

(*Continued*)

Host	Letter	Number	p (days)	d (AU)	M (Earths)	R (Earths)	Temperature (K)	M_s (Suns)	R_s (Suns)
Kepler-200	b	2	8.59	0.08	0	2.128	5678		0.94
Kepler-200	c	2	10.22	0.09	0	1.5904	5678		0.94
Kepler-201	b	2	25.67	0.175	0	2.4528	6065	1.17	1.23
Kepler-201	c	2	151.88	0.573	0	2.8448	6065	1.17	1.23
Kepler-202	b	2	4.07	0.045	0	1.624	4668		0.67
Kepler-202	c	2	16.28	0.113	0	1.848	4668		0.67
Kepler-203	b	3	3.16	0.043	0	2.5648	5821	0.98	1.11
Kepler-203	c	3	5.37	0.061	0	2.464	5821	0.98	1.11
Kepler-203	d	3	11.33	0.1	0	1.4336	5821	0.98	1.11
Kepler-204	b	2	14.4	0.117	0	2.5312	5812	0.96	1.24
Kepler-204	c	2	25.66	0.173	0	1.792	5812	0.96	1.24
Kepler-205	b	2	2.76	0.032	0	1.512	4321		0.55
Kepler-205	c	2	20.31	0.122	0	1.6352	4321		0.55
Kepler-206	b	3	7.78	0.078	0	1.1984	5764	0.94	1.19
Kepler-206	c	3	13.14	0.111	0	1.7696	5764	0.94	1.19
Kepler-206	d	3	23.44	0.163	0	1.1872	5764	0.94	1.19
Kepler-207	b	3	1.61	0.029	0	1.568	5920		1.59
Kepler-207	c	3	3.07	0.044	0	1.5008	5920		1.59
Kepler-207	d	3	5.87	0.068	0	3.304	5920		1.59
Kepler-208	b	4	4.23	0.054	0	1.624	6092	1.03	1.31
Kepler-208	c	4	7.47	0.079	0	1.3888	6092	1.03	1.31
Kepler-208	d	4	11.13	0.103	0	1.1984	6092	1.03	1.31
Kepler-208	e	4	16.26	0.132	0	1.4784	6092	1.03	1.31
Kepler-209	b	2	16.09	0.122	0	2.2624	5513		0.94
Kepler-209	c	2	41.75	0.231	0	3.1024	5513		0.94
Kepler-21	b	1	2.79	0.04251	10.494	1.6352	6131	1.34	1.86
Kepler-210	b	2	2.45	0.032	0	2.9344	4559		0.65
Kepler-210	c	2	7.97	0.07	0	3.6176	4559		0.65
Kepler-211	b	2	4.14	0.048	0	1.2544	5123	0.97	0.82
Kepler-211	c	2	6.04	0.062	0	1.288	5123	0.97	0.82
Kepler-212	b	2	16.26	0.133	0	1.0864	5852	1.16	1.46
Kepler-212	c	2	31.81	0.207	0	2.7328	5852	1.16	1.46
Kepler-213	b	2	2.46	0.036	0	1.624	5696	0.94	1.2
Kepler-213	c	2	4.82	0.057	0	2.3408	5696	0.94	1.2
Kepler-214	b	2	15.66	0.13	0	2.6096	6169		1.35
Kepler-214	c	2	28.78	0.194	0	2.128	6169		1.35
Kepler-215	b	4	9.36	0.084	0	1.624	5739	0.77	1.03
Kepler-215	c	4	14.67	0.113	0	1.7696	5739	0.77	1.03
Kepler-215	d	4	30.86	0.185	0	2.3856	5739	0.77	1.03
Kepler-215	e	4	68.16	0.314	0	1.7472	5739	0.77	1.03
Kepler-216	b	2	7.69	0.079	0	2.352	6091		1.26
Kepler-216	c	2	17.41	0.136	0	3.0352	6091		1.26
Kepler-217	b	3	5.37	0.065	0	2.2288	6171		1.8
Kepler-217	c	3	8.59	0.089	0	1.848	6171		1.8
Kepler-218	b	3	3.62	0.046	0	1.4784	5502		1.06
Kepler-218	c	3	44.7	0.248	0	3.136	5502		1.06

Kepler-219	b	3	4.59	0.057	0	2.9456	5786		1.49
Kepler-219	c	3	22.71	0.165	0	3.5728	5786		1.49
Kepler-219	d	3	47.9	0.272	0	2.8112	5786		1.49
Kepler-22	b	1	289.86	0.849	35.934	2.3744	5518	0.97	0.98
Kepler-220	b	4	4.16	0.046	0	0.8064	4632		0.67
Kepler-220	c	4	9.03	0.076	0	1.568	4632		0.67
Kepler-220	d	4	28.12	0.163	0	0.9744	4632		0.67
Kepler-220	e	4	45.9	0.226	0	1.3328	4632		0.67
Kepler-221	b	4	2.8	0.037	0	1.7136	5243	0.72	0.82
Kepler-221	c	4	5.69	0.059	0	2.9232	5243	0.72	0.82
Kepler-221	d	4	10.04	0.087	0	2.7328	5243	0.72	0.82
Kepler-221	e	4	18.37	0.13	0	2.632	5243	0.72	0.82
Kepler-222	b	3	3.94	0.048	0	3.1584	5433		0.87
Kepler-222	c	3	10.09	0.091	0	4.6368	5433		0.87
Kepler-222	d	3	28.08	0.18	0	3.6848	5433		0.87
Kepler-223	b	4	7.38	0.073	0	1.6912	5829		1.02
Kepler-223	c	4	9.85	0.088	0	1.9936	5829		1.02
Kepler-223	d	4	14.79	0.116	0	2.9792	5829		1.02
Kepler-223	e	4	19.72	0.14	0	2.3968	5829		1.02
Kepler-224	b	4	3.13	0.038	0	1.3888	5018	0.74	0.68
Kepler-224	c	4	5.93	0.058	0	3.1136	5018	0.74	0.68
Kepler-224	d	4	11.35	0.089	0	2.296	5018	0.74	0.68
Kepler-224	e	4	18.64	0.124	0	1.9712	5018	0.74	0.68
Kepler-225	b	2	6.74	0.056	0	1.1984	3682		0.48
Kepler-225	c	2	18.79	0.111	0	1.8368	3682		0.48
Kepler-226	b	3	3.94	0.047	0	1.5456	5571	0.86	0.8
Kepler-226	c	3	5.35	0.058	0	2.2736	5571	0.86	0.8
Kepler-226	d	3	8.11	0.076	0	1.2208	5571	0.86	0.8
Kepler-227	b	2	9.49	0.09	0	3.1024	5854		1.09
Kepler-227	c	2	54.42	0.29	0	3.0352	5854		1.09
Kepler-228	b	3	2.57	0.038	0	1.5232	6043		1.01
Kepler-228	c	3	4.13	0.052	0	2.6992	6043		1.01
Kepler-228	d	3	11.09	0.101	0	4.032	6043		1.01
Kepler-229	b	3	6.25	0.062	0	2.1952	5120		0.73
Kepler-229	c	3	16.07	0.117	0	4.9168	5120		0.73
Kepler-229	d	3	41.19	0.22	0	3.8416	5120		0.73
Kepler-23	b	3	7.11	0.075	254.4	1.904	5828	1.11	1.55
Kepler-23	c	3	10.74	0.099	858.6	3.192	5828	1.11	1.55
Kepler-23	d	3	15.27	0.124	0	2.1952	5828	1.11	1.55
Kepler-230	b	2	32.63	0.191	0	4.256	5588		0.82
Kepler-230	c	2	91.77	0.38	0	2.0384	5588		0.82
Kepler-231	b	2	10.07	0.074	0	1.7248	3767	0.58	0.49
Kepler-231	c	2	19.27	0.114	0	1.9264	3767	0.58	0.49
Kepler-232	b	2	4.43	0.054	0	3.08	5847		0.97
Kepler-232	c	2	11.38	0.101	0	3.8304	5847		0.97
Kepler-233	b	2	8.47	0.077	0	2.4304	5360		0.76
Kepler-233	c	2	60.42	0.287	0	2.7104	5360		0.76
Kepler-234	b	2	2.71	0.04	0	3.696	6224		1.11
Kepler-234	c	2	7.21	0.077	0	3.5056	6224		1.11

(Continued)

(*Continued*)

Host	Letter	Number	p (days)	d (AU)	M (Earths)	R (Earths)	Temperature (K)	M_s (Suns)	R_s (Suns)
Kepler-235	b	4	3.34	0.037	0	2.2288	4255	0.59	0.55
Kepler-235	c	4	7.82	0.065	0	1.2768	4255	0.59	0.55
Kepler-235	d	4	20.06	0.122	0	2.0496	4255	0.59	0.55
Kepler-235	e	4	46.18	0.213	0	2.2176	4255	0.59	0.55
Kepler-236	b	2	8.3	0.065	0	1.568	3750	0.56	0.51
Kepler-236	c	2	23.97	0.132	0	1.9936	3750	0.56	0.51
Kepler-237	b	2	4.72	0.05	0	1.4112	4861	0.7	0.72
Kepler-237	c	2	8.1	0.071	0	2.0832	4861	0.7	0.72
Kepler-238	b	5	2.09	0.034	0	1.7248	5751	1.06	1.43
Kepler-238	c	5	6.16	0.069	0	2.3856	5751	1.06	1.43
Kepler-238	d	5	13.23	0.115	0	3.0688	5751	1.06	1.43
Kepler-239	b	2	11.76	0.095	0	2.3296	4914	0.74	0.76
Kepler-239	c	2	56.23	0.268	0	2.5088	4914	0.74	0.76
Kepler-24	b	4	8.15	0.08	508.8	2.3968	5897	1.03	1.29
Kepler-24	c	4	12.33	0.106	508.8	2.8	5897	1.03	1.29
Kepler-24	d	4	4.24	0.051	0	1.6688	5897	1.03	1.29
Kepler-24	e	4	19	0.138	0	2.7776	5897	1.03	1.29
Kepler-240	b	2	4.14	0.048	0	1.3664	4985		0.74
Kepler-240	c	2	7.95	0.074	0	2.1952	4985		0.74
Kepler-241	b	2	12.72	0.094	0	2.3296	4699		0.67
Kepler-241	c	2	36.07	0.189	0	2.5648	4699		0.67
Kepler-242	b	2	8.2	0.075	0	2.6096	5020		0.85
Kepler-242	c	2	14.5	0.109	0	1.9936	5020		0.85
Kepler-243	b	2	5.72	0.062	0	2.4528	5228	0.89	0.84
Kepler-243	c	2	20.03	0.142	0	1.9936	5228	0.89	0.84
Kepler-244	b	3	4.31	0.05	0	2.7552	5554		0.8
Kepler-244	c	3	9.77	0.087	0	2.0496	5554		0.8
Kepler-244	d	3	20.05	0.14	0	2.3072	5554		0.8
Kepler-245	b	4	7.49	0.071	0	2.5648	5100	0.8	0.8
Kepler-245	c	4	17.46	0.124	0	2.1728	5100	0.8	0.8
Kepler-245	d	4	36.28	0.202	0	3.024	5100	0.8	0.8
Kepler-246	b	2	4.6	0.052	0	2.3408	5206	0.86	0.83
Kepler-246	c	2	11.19	0.095	0	1.5008	5206	0.86	0.83
Kepler-247	b	3	3.34	0.042	0	1.6352	5100		0.77
Kepler-247	c	3	9.44	0.084	0	4.088	5100		0.77
Kepler-247	d	3	20.48	0.14	0	3.9424	5100		0.77
Kepler-248	b	2	6.31	0.066	0	3.0128	5190		0.83
Kepler-248	c	2	16.24	0.123	0	4.0656	5190		0.83
Kepler-249	b	3	3.31	0.035	0	1.0864	3568		0.48
Kepler-249	c	3	7.11	0.058	0	1.512	3568		0.48
Kepler-249	d	3	15.37	0.097	0	1.568	3568		0.48
Kepler-25	b	3	6.24	0.068	4038.6	2.5984	6270	1.19	1.31
Kepler-25	c	3	12.72	0.11	1322.88	4.4912	6270	1.19	1.31
Kepler-250	b	3	4.15	0.048	0	1.1312	5160	0.8	0.81
Kepler-250	c	3	7.16	0.069	0	2.2736	5160	0.8	0.81
Kepler-250	d	3	17.65	0.127	0	2.1728	5160	0.8	0.81

A-8

Kepler-251	b	4	4.79	0.053	0	1.3328	5526	0.91	0.89
Kepler-251	c	4	16.51	0.122	0	2.7664	5526	0.91	0.89
Kepler-251	d	4	30.13	0.182	0	2.7664	5526	0.91	0.89
Kepler-251	e	4	99.64	0.404	0	2.7664	5526	0.91	0.89
Kepler-252	b	2	6.67	0.058	0	1.232	4208	0.52	0.55
Kepler-252	c	2	10.85	0.08	0	2.1504	4208	0.52	0.55
Kepler-253	b	3	3.78	0.046	0	1.624	5208		0.79
Kepler-253	c	3	10.28	0.089	0	2.6432	5208		0.79
Kepler-253	d	3	18.12	0.13	0	3.1696	5208		0.79
Kepler-254	b	3	5.83	0.064	0	3.864	5957		0.91
Kepler-254	c	3	12.41	0.105	0	2.1504	5957		0.91
Kepler-254	d	3	18.75	0.139	0	2.4976	5957		0.91
Kepler-255	b	3	5.71	0.063	0	1.5456	5573	0.97	0.93
Kepler-255	c	3	9.95	0.092	0	2.9904	5573	0.97	0.93
Kepler-256	b	4	1.62	0.027	0	1.5904	5551	1.02	1.3
Kepler-256	c	4	3.39	0.045	0	2.1504	5551	1.02	1.3
Kepler-256	d	4	5.84	0.064	0	2.4752	5551	1.02	1.3
Kepler-256	e	4	10.68	0.096	0	2.352	5551	1.02	1.3
Kepler-257	b	3	2.38	0.034	0	2.6096	5180		1.04
Kepler-257	c	3	6.58	0.066	0	5.4096	5180		1.04
Kepler-257	d	3	24.66	0.16	0	4.9504	5180		1.04
Kepler-258	b	2	13.2	0.103	0	4.0544	4942	0.8	0.92
Kepler-258	c	2	33.65	0.193	0	3.6064	4942	0.8	0.92
Kepler-259	b	2	8.12	0.079	0	2.8	5938		0.9
Kepler-259	c	2	36.92	0.217	0	2.6992	5938		0.9
Kepler-26	d	4	3.54	0.039	0	1.064	3914	0.54	0.51
Kepler-26	e	4	46.83	0.22	0	2.408	3914	0.54	0.51
Kepler-260	b	2	8.19	0.075	0	2.0048	5250	0.88	0.86
Kepler-260	c	2	76.05	0.332	0	1.736	5250	0.88	0.86
Kepler-261	b	2	10.38	0.088	0	2.1728	5098	0.87	0.79
Kepler-261	c	2	24.57	0.156	0	1.9936	5098	0.87	0.79
Kepler-262	b	2	13.06	0.108	0	1.3552	5841		0.88
Kepler-262	c	2	21.85	0.152	0	1.6352	5841		0.88
Kepler-263	b	2	16.57	0.12	0	2.6656	5265		0.79
Kepler-263	c	2	47.33	0.242	0	2.464	5265		0.79
Kepler-264	b	2	40.81	0.249	0	3.3264	6158	1.32	1.55
Kepler-264	c	2	140.1	0.566	0	2.8224	6158	1.32	1.55
Kepler-265	b	4	6.85	0.069	0	1.8592	5835	1.03	1.1
Kepler-265	c	4	17.03	0.127	0	2.632	5835	1.03	1.1
Kepler-265	d	4	43.13	0.236	0	2.4864	5835	1.03	1.1
Kepler-265	e	4	67.83	0.319	0	2.5872	5835	1.03	1.1
Kepler-266	b	2	6.62	0.071	0	2.4752	5885	0.99	1.03
Kepler-266	c	2	107.72	0.457	0	3.8864	5885	0.99	1.03
Kepler-267	b	3	3.35	0.037	0	1.9824	4258	0.56	0.56
Kepler-267	c	3	6.88	0.06	0	2.128	4258	0.56	0.56
Kepler-267	d	3	28.46	0.154	0	2.2736	4258	0.56	0.56
Kepler-268	b	2	25.93	0.18	0	2.5424	6081		1.27
Kepler-268	c	2	83.45	0.391	0	3.3824	6081		1.27
Kepler-269	b	2	5.33	0.061	0	2.464	5847	0.98	0.96

(Continued)

(*Continued*)

Host	Letter	Number	p (days)	d (AU)	M (Earths)	R (Earths)	Temperature (K)	M_s (Suns)	R_s (Suns)
Kepler-269	c	2	8.13	0.081	0	1.6912	5847	0.98	0.96
Kepler-27	b	2	15.33	0.118	2896.98	3.9984	5400	0.65	0.59
Kepler-27	c	2	31.33	0.191	4388.4	4.8944	5400	0.65	0.59
Kepler-270	b	2	11.48	0.107	0	2.0048	6067		1.46
Kepler-270	c	2	25.26	0.18	0	1.7696	6067		1.46
Kepler-271	b	3	5.22	0.056	0	0.8848	5524		0.88
Kepler-271	c	3	7.41	0.071	0	0.9968	5524		0.88
Kepler-272	b	3	2.97	0.038	0	1.4336	5297	0.79	0.93
Kepler-272	c	3	6.06	0.061	0	1.792	5297	0.79	0.93
Kepler-272	d	3	10.94	0.091	0	2.2512	5297	0.79	0.93
Kepler-273	b	2	2.94	0.037	0	1.5008	5626		0.81
Kepler-273	c	2	8.01	0.073	0	1.9824	5626		0.81
Kepler-274	b	2	11.63	0.101	0	1.5344	6023		1.01
Kepler-274	c	2	33.2	0.204	0	1.8368	6023		1.01
Kepler-275	b	3	10.3	0.098	0	2.3408	6165	1.24	1.38
Kepler-275	c	3	16.09	0.132	0	3.3824	6165	1.24	1.38
Kepler-275	d	3	35.68	0.224	0	3.3264	6165	1.24	1.38
Kepler-276	b	3	14.13	0.119	0	2.8672	6105	1.1	1.05
Kepler-278	b	2	30.16	0.207	0	4.0656	4991		2.94
Kepler-278	c	2	51.08	0.294	0	3.584	4991		2.94
Kepler-279	b	3	12.31	0.112	0	3.6176	6363	1.1	1.75
Kepler-28	b	2	5.91	0.062	480.18	3.5952	4590	0.75	0.7
Kepler-28	c	2	8.99	0.081	432.48	3.3936	4590	0.75	0.7
Kepler-280	b	2	2.14	0.032	0	1.4448	5744	0.91	0.89
Kepler-280	c	2	4.81	0.056	0	2.0048	5744	0.91	0.89
Kepler-281	b	2	14.65	0.117	0	2.8224	5723	0.98	0.9
Kepler-281	c	2	36.34	0.215	0	5.3088	5723	0.98	0.9
Kepler-282	b	4	9.22	0.082	0	1.008	5602	0.97	0.9
Kepler-282	c	4	13.64	0.106	0	1.1984	5602	0.97	0.9
Kepler-283	b	2	11.01	0.082	0	2.128	4351		0.57
Kepler-283	c	2	92.74	0.341	0	1.8144	4351		0.57
Kepler-284	b	2	12.7	0.104	0	2.24	5615		0.81
Kepler-284	c	2	37.51	0.213	0	2.6096	5615		0.81
Kepler-285	b	2	2.63	0.036	0	1.344	5411	0.85	0.81
Kepler-285	c	2	6.19	0.064	0	1.12	5411	0.85	0.81
Kepler-286	b	4	1.8	0.027	0	1.2432	5580		0.86
Kepler-286	c	4	3.47	0.042	0	1.3664	5580		0.86
Kepler-286	d	4	5.91	0.061	0	1.3328	5580		0.86
Kepler-286	e	4	29.22	0.176	0	1.7696	5580		0.86
Kepler-287	b	2	20.34	0.145	0	2.3296	5806		1.03
Kepler-287	c	2	44.85	0.246	0	3.2592	5806		1.03
Kepler-288	b	3	6.1	0.065	0	1.6688	5918	0.89	1.09
Kepler-288	c	3	19.31	0.14	0	2.8448	5918	0.89	1.09
Kepler-288	d	3	56.63	0.287	0	2.6656	5918	0.89	1.09
Kepler-289	b	3	34.55	0.21	7.314	2.1504	5990	1.08	1
Kepler-289	c	3	125.85	0.51	133.56	11.5808	5990	1.08	1

Kepler-289	d	3	66.06	0.33	4.134	2.6768	5990	1.08	1
Kepler-290	b	2	14.59	0.11	0	2.2512	5147		0.74
Kepler-290	c	2	36.77	0.205	0	2.6992	5147		0.74
Kepler-291	b	2	3.55	0.047	0	2.1616	6002	1.03	1.02
Kepler-291	c	2	5.7	0.065	0	1.8816	6002	1.03	1.02
Kepler-292	b	5	2.58	0.035	0	1.3216	5299	0.88	0.83
Kepler-292	c	5	3.72	0.045	0	1.4672	5299	0.88	0.83
Kepler-292	d	5	7.06	0.068	0	2.2288	5299	0.88	0.83
Kepler-292	e	5	11.98	0.097	0	2.6656	5299	0.88	0.83
Kepler-292	f	5	20.83	0.141	0	2.352	5299	0.88	0.83
Kepler-293	b	2	19.25	0.144	0	3.0688	5804	1.01	0.96
Kepler-293	c	2	54.16	0.286	0	3.8304	5804	1.01	0.96
Kepler-294	b	2	3.7	0.048	0	1.7696	5913		0.98
Kepler-294	c	2	6.63	0.071	0	2.7104	5913		0.98
Kepler-295	b	3	12.65	0.099	0	1.2208	5603		0.9
Kepler-295	c	3	21.53	0.142	0	1.1648	5603		0.9
Kepler-295	d	3	33.88	0.192	0	1.3552	5603		0.9
Kepler-296	b	5	10.86	0.079	0	1.6128	3740	0.5	0.48
Kepler-296	c	5	5.84	0.0521	0	1.9936	3740	0.5	0.48
Kepler-296	d	5	19.85	0.118	0	2.0832	3740	0.5	0.48
Kepler-296	e	5	34.14	0.169	0	1.5232	3740	0.5	0.48
Kepler-296	f	5	63.34	0.255	0	1.8032	3740	0.5	0.48
Kepler-297	b	2	38.87	0.217	0	2.8672	5619		0.92
Kepler-297	c	2	74.92	0.336	0	6.5296	5619		0.92
Kepler-298	b	3	10.48	0.08	0	1.96	4465	0.65	0.58
Kepler-298	c	3	22.93	0.136	0	1.9264	4465	0.65	0.58
Kepler-298	d	3	77.47	0.305	0	2.4976	4465	0.65	0.58
Kepler-299	b	4	2.93	0.04	0	1.3216	5617	0.97	1.03
Kepler-299	c	4	6.89	0.07	0	2.6432	5617	0.97	1.03
Kepler-299	d	4	15.05	0.118	0	1.8592	5617	0.97	1.03
Kepler-299	e	4	38.29	0.22	0	1.8704	5617	0.97	1.03
Kepler-30	b	3	29.33	0.18	11.448	3.8976	5498	0.99	0.95
Kepler-30	c	3	60.32	0.3	639.18	12.2864	5498	0.99	0.95
Kepler-30	d	3	143.34	0.5	23.214	8.792	5498	0.99	0.95
Kepler-300	b	2	10.45	0.094	0	1.6688	5986	0.94	0.9
Kepler-300	c	2	40.71	0.232	0	2.2512	5986	0.94	0.9
Kepler-301	b	3	2.51	0.036	0	1.344	5815	0.91	0.9
Kepler-301	c	3	5.42	0.06	0	1.344	5815	0.91	0.9
Kepler-301	d	3	13.75	0.112	0	1.7472	5815	0.91	0.9
Kepler-302	b	2	30.18	0.193	0	4.0544	5740		1.22
Kepler-302	c	2	127.28	0.503	0	12.4432	5740		1.22
Kepler-303	b	2	1.94	0.024	0	0.8848	3944	0.59	0.48
Kepler-303	c	2	7.06	0.057	0	1.1424	3944	0.59	0.48
Kepler-304	b	4	3.3	0.039	0	2.856	4731	0.8	0.69
Kepler-304	c	4	5.32	0.054	0	2.1728	4731	0.8	0.69
Kepler-304	d	4	9.65	0.08	0	2.744	4731	0.8	0.69
Kepler-305	d	3	16.74	0.121	0	2.7104	5100	0.76	0.79
Kepler-306	b	4	4.65	0.05	0	1.5232	4954	0.82	0.72
Kepler-306	c	4	7.24	0.067	0	2.352	4954	0.82	0.72

(Continued)

A-11

(*Continued*)

Host	Letter	Number	p (days)	d (AU)	M (Earths)	R (Earths)	Temperature (K)	M_s (Suns)	R_s (Suns)
Kepler-306	d	4	17.33	0.12	0	2.464	4954	0.82	0.72
Kepler-306	e	4	44.84	0.227	0	2.2736	4954	0.82	0.72
Kepler-308	b	2	9.69	0.09	0	2.1168	5895		0.94
Kepler-308	c	2	15.38	0.123	0	2.1616	5895		0.94
Kepler-309	b	2	5.92	0.059	0	1.5568	4713		0.72
Kepler-309	c	2	105.36	0.401	0	2.5088	4713		0.72
Kepler-31	b	3	20.86	0.16	0	5.4992	6340	1.21	1.22
Kepler-31	c	3	42.63	0.26	1494.6	5.2976	6340	1.21	1.22
Kepler-31	d	3	87.65	0.4	2162.4	3.8976	6340	1.21	1.22
Kepler-310	b	3	13.93	0.111	0	1.1872	5797	0.85	0.88
Kepler-310	c	3	56.48	0.281	0	3.3824	5797	0.85	0.88
Kepler-310	d	3	92.88	0.392	0	2.464	5797	0.85	0.88
Kepler-311	b	2	9.18	0.087	0	1.7024	5905	1.06	1.19
Kepler-311	c	2	19.74	0.145	0	1.4336	5905	1.06	1.19
Kepler-312	b	2	1.77	0.031	0	1.288	6115	1.23	1.47
Kepler-312	c	2	19.75	0.153	0	3.1472	6115	1.23	1.47
Kepler-313	b	2	14.97	0.125	0	2.5312	5727	0.9	1.54
Kepler-313	c	2	32.27	0.208	0	2.5648	5727	0.9	1.54
Kepler-314	b	2	2.46	0.035	0	0.8288	5378	1.02	0.95
Kepler-314	c	2	5.96	0.064	0	2.9008	5378	1.02	0.95
Kepler-315	b	2	96.1	0.402	0	3.7632	5796	0.78	1.04
Kepler-315	c	2	265.47	0.791	0	4.144	5796	0.78	1.04
Kepler-316	b	2	2.24	0.027	0	1.064	4204	0.53	0.52
Kepler-316	c	2	6.83	0.058	0	1.1536	4204	0.53	0.52
Kepler-317	b	2	5.52	0.061	0	2.0832	5497	0.95	0.94
Kepler-317	c	2	8.78	0.083	0	1.7136	5497	0.95	0.94
Kepler-318	b	2	4.66	0.056	0	4.704	5746	1.05	1.19
Kepler-318	c	2	11.82	0.105	0	3.6736	5746	1.05	1.19
Kepler-319	b	3	4.36	0.051	0	1.624	5526	1.29	0.9
Kepler-319	c	3	6.94	0.069	0	2.632	5526	1.29	0.9
Kepler-319	d	3	31.78	0.191	0	2.2848	5526	1.29	0.9
Kepler-32	b	5	5.9	0.05	1303.8	2.1952	3900	0.58	0.53
Kepler-32	c	5	8.75	0.09	159	1.9936	3900	0.58	0.53
Kepler-32	d	5	22.78	0.13	0	2.6992	3900	0.58	0.53
Kepler-32	e	5	2.9	0.033	0	1.5008	3900	0.58	0.53
Kepler-32	f	5	0.74	0.013	0	0.8176	3900	0.58	0.53
Kepler-320	b	2	8.37	0.085	0	1.1424	6435		1.11
Kepler-320	c	2	17.93	0.142	0	1.3664	6435		1.11
Kepler-321	b	2	4.92	0.057	0	1.7696	5740	1.01	1.19
Kepler-321	c	2	13.09	0.11	0	2.3184	5740	1.01	1.19
Kepler-322	b	2	1.65	0.027	0	1.008	5388	0.91	0.89
Kepler-322	c	2	4.34	0.051	0	1.6688	5388	0.91	0.89
Kepler-323	b	2	1.68	0.028	0	1.4336	5987	1.09	1.18
Kepler-323	c	2	3.55	0.046	0	1.624	5987	1.09	1.18
Kepler-324	b	2	4.39	0.05	0	1.1424	5194	0.86	0.84
Kepler-324	c	2	51.81	0.26	0	3.1472	5194	0.86	0.84

Kepler-325	b	3	4.54	0.053	0	2.912	5752	0.87	1
Kepler-325	c	3	12.76	0.105	0	2.5424	5752	0.87	1
Kepler-325	d	3	38.72	0.22	0	2.7888	5752	0.87	1
Kepler-326	b	3	2.25	0.032	0	1.5232	5105	0.98	0.8
Kepler-326	c	3	4.58	0.051	0	1.4	5105	0.98	0.8
Kepler-326	d	3	6.77	0.066	0	1.2096	5105	0.98	0.8
Kepler-327	b	3	2.55	0.029	0	1.1088	3799	0.55	0.49
Kepler-327	c	3	5.21	0.047	0	1.0304	3799	0.55	0.49
Kepler-327	d	3	13.97	0.09	0	1.7248	3799	0.55	0.49
Kepler-329	b	2	7.42	0.061	0	1.4	4257	0.53	0.52
Kepler-329	c	2	18.68	0.113	0	1.9264	4257	0.53	0.52
Kepler-33	b	5	5.67	0.0677	0	1.736	5904	1.29	1.82
Kepler-33	c	5	13.18	0.1189	0	3.192	5904	1.29	1.82
Kepler-33	d	5	21.78	0.1662	0	5.3424	5904	1.29	1.82
Kepler-33	e	5	31.78	0.2138	0	4.0208	5904	1.29	1.82
Kepler-33	f	5	41.03	0.2535	0	4.4576	5904	1.29	1.82
Kepler-330	b	2	8.26	0.075	0	1.344	5117	0.78	0.72
Kepler-330	c	2	15.96	0.116	0	1.9488	5117	0.78	0.72
Kepler-331	b	3	8.46	0.065	0	1.8144	4347	0.51	0.49
Kepler-331	c	3	17.28	0.105	0	1.8368	4347	0.51	0.49
Kepler-331	d	3	32.13	0.159	0	1.6352	4347	0.51	0.49
Kepler-332	b	3	7.63	0.07	0	1.1648	4955	0.8	0.72
Kepler-332	c	3	16	0.114	0	1.0864	4955	0.8	0.72
Kepler-332	d	3	34.21	0.189	0	1.176	4955	0.8	0.72
Kepler-333	b	2	12.55	0.087	0	1.3216	4259	0.54	0.53
Kepler-333	c	2	24.09	0.135	0	1.1088	4259	0.54	0.53
Kepler-334	b	3	5.47	0.061	0	1.12	5828	1	1.07
Kepler-334	c	3	12.76	0.107	0	1.4336	5828	1	1.07
Kepler-334	d	3	25.1	0.168	0	1.4112	5828	1	1.07
Kepler-335	b	2	6.56	0.075	0	3.3824	5877	0.99	1.85
Kepler-335	c	2	67.84	0.356	0	3.0688	5877	0.99	1.85
Kepler-336	b	3	2.02	0.033	0	1.0192	5867	0.89	1.3
Kepler-336	c	3	9.6	0.092	0	2.0944	5867	0.89	1.3
Kepler-336	d	3	20.68	0.154	0	2.3632	5867	0.89	1.3
Kepler-337	b	2	3.29	0.045	0	1.5344	5684	0.96	1.76
Kepler-337	c	2	9.69	0.093	0	2.0496	5684	0.96	1.76
Kepler-338	b	4	13.73	0.117	0	2.4416	5923	1.1	1.74
Kepler-338	c	4	24.31	0.172	0	2.3408	5923	1.1	1.74
Kepler-338	d	4	44.43	0.257	0	3.0016	5923	1.1	1.74
Kepler-339	b	3	4.98	0.055	0	1.4224	5631	0.84	0.8
Kepler-339	c	3	6.99	0.069	0	1.1536	5631	0.84	0.8
Kepler-339	d	3	10.56	0.091	0	1.1648	5631	0.84	0.8
Kepler-34	b	1	288.82	1.0896	69.96	8.5568	5913	1.05	1.16
Kepler-340	b	2	14.84	0.134	0	2.5312	6620	2.11	1.85
Kepler-340	c	2	22.82	0.178	0	3.3712	6620	2.11	1.85
Kepler-341	b	4	5.2	0.06	0	1.176	6012	0.94	1.02
Kepler-341	c	4	8.01	0.08	0	1.7024	6012	0.94	1.02
Kepler-341	d	4	27.67	0.182	0	1.848	6012	0.94	1.02
Kepler-341	e	4	42.47	0.242	0	1.9936	6012	0.94	1.02

(Continued)

(*Continued*)

Host	Letter	Number	p (days)	d (AU)	M (Earths)	R (Earths)	Temperature (K)	M_s (Suns)	R_s (Suns)
Kepler-342	b	4	15.17	0.128	0	2.2512	6175	1.13	1.47
Kepler-342	c	4	26.23	0.185	0	1.96	6175	1.13	1.47
Kepler-342	d	4	39.46	0.242	0	2.4864	6175	1.13	1.47
Kepler-343	b	2	8.97	0.088	0	2.408	5807	1.04	1.43
Kepler-343	c	2	23.22	0.167	0	2.016	5807	1.04	1.43
Kepler-344	b	2	21.96	0.153	0	2.6096	5774	0.9	0.98
Kepler-344	c	2	125.6	0.488	0	2.9456	5774	0.9	0.98
Kepler-345	b	2	7.42	0.066	0	0.7392	4504	0.59	0.62
Kepler-345	c	2	9.39	0.077	0	1.1984	4504	0.59	0.62
Kepler-346	b	2	6.51	0.071	0	2.6544	6033	0.97	1.02
Kepler-346	c	2	23.85	0.168	0	3.0688	6033	0.97	1.02
Kepler-347	b	2	12.8	0.11	0	1.9712	6088	1.04	1
Kepler-347	c	2	27.32	0.183	0	1.7472	6088	1.04	1
Kepler-348	b	2	7.06	0.076	0	1.5232	6177	1.15	1.36
Kepler-348	c	2	17.27	0.138	0	1.3328	6177	1.15	1.36
Kepler-349	b	2	5.93	0.065	0	1.904	5956	0.97	0.93
Kepler-349	c	2	12.25	0.105	0	1.96	5956	0.97	0.93
Kepler-35	b	1	131.46	0.60347	40.386	8.1536	5606	0.89	1.03
Kepler-350	b	3	11.19	0.104	0	1.848	6186	1	1.53
Kepler-351	b	3	37.05	0.214	0	3.0576	5643	0.89	0.85
Kepler-351	c	3	57.25	0.287	0	3.192	5643	0.89	0.85
Kepler-352	b	2	10.06	0.085	0	0.8624	5212	0.79	0.78
Kepler-352	c	2	16.33	0.118	0	1.2432	5212	0.79	0.78
Kepler-353	b	2	5.8	0.051	0	0.8848	3903	0.54	0.5
Kepler-353	c	2	8.41	0.065	0	1.3776	3903	0.54	0.5
Kepler-354	b	3	5.48	0.054	0	1.8368	4648	0.65	0.67
Kepler-354	c	3	16.93	0.115	0	1.3104	4648	0.65	0.67
Kepler-354	d	3	24.21	0.146	0	1.2432	4648	0.65	0.67
Kepler-355	b	2	11.03	0.102	0	1.456	6184	1.05	1.07
Kepler-355	c	2	25.76	0.179	0	2.7104	6184	1.05	1.07
Kepler-356	b	2	4.61	0.057	0	1.568	6133	0.97	1.33
Kepler-356	c	2	13.12	0.115	0	1.8032	6133	0.97	1.33
Kepler-357	b	3	6.48	0.063	0	1.8368	5036	0.78	0.83
Kepler-357	c	3	16.86	0.12	0	2.6656	5036	0.78	0.83
Kepler-357	d	3	49.5	0.246	0	3.4272	5036	0.78	0.83
Kepler-358	b	2	34.06	0.21	0	2.7216	5908	0.95	0.95
Kepler-358	c	2	83.49	0.381	0	2.8448	5908	0.95	0.95
Kepler-359	b	3	25.56	0.178	0	3.528	6248	1.07	1.09
Kepler-359	c	3	57.69	0.307	0	4.3008	6248	1.07	1.09
Kepler-359	d	3	77.1	0.372	0	4.0096	6248	1.07	1.09
Kepler-36	b	2	13.84	0.1153	4.452	1.4896	5911	1.07	1.63
Kepler-36	c	2	16.24	0.1283	7.95	3.6736	5911	1.07	1.63
Kepler-360	b	2	3.29	0.044	0	1.6464	6053	0.95	1.06
Kepler-360	c	2	7.19	0.075	0	2.0944	6053	0.95	1.06
Kepler-361	b	2	8.49	0.086	0	1.4448	6169	1.07	1.34
Kepler-361	c	2	55.19	0.3	0	2.52	6169	1.07	1.34

Kepler-362	b	2	10.33	0.087	0	0.8848	5788	0.77	0.72
Kepler-362	c	2	37.87	0.207	0	1.4448	5788	0.77	0.72
Kepler-363	b	3	3.61	0.048	0	1.1536	5593	1.23	1.49
Kepler-363	c	3	7.54	0.079	0	1.6912	5593	1.23	1.49
Kepler-363	d	3	11.93	0.107	0	2.0496	5593	1.23	1.49
Kepler-364	b	2	25.75	0.178	0	1.5456	6108	1.2	1.28
Kepler-364	c	2	59.98	0.312	0	2.1504	6108	1.2	1.28
Kepler-365	b	2	10.66	0.098	0	2.0384	6012	0.99	1.05
Kepler-365	c	2	17.78	0.137	0	1.6352	6012	0.99	1.05
Kepler-366	b	2	3.28	0.045	0	1.456	6209	1.05	1.05
Kepler-366	c	2	12.52	0.11	0	1.792	6209	1.05	1.05
Kepler-367	b	2	37.82	0.201	0	1.2992	4710	0.75	0.69
Kepler-367	c	2	53.58	0.253	0	1.1984	4710	0.75	0.69
Kepler-368	b	2	26.85	0.186	0	3.2592	5502	0.71	2.02
Kepler-368	c	2	72.38	0.36	0	3.8752	5502	0.71	2.02
Kepler-369	b	2	2.73	0.03	0	1.1312	3591	0.54	0.47
Kepler-369	c	2	14.87	0.094	0	1.4112	3591	0.54	0.47
Kepler-37	b	4	13.37	0.1003	0	0.3024	5417	0.8	0.77
Kepler-37	c	4	21.3	0.1368	0	0.7392	5417	0.8	0.77
Kepler-37	d	4	39.79	0.2076	0	1.9936	5417	0.8	0.77
Kepler-370	b	2	4.58	0.054	0	1.5904	5852	0.94	0.9
Kepler-370	c	2	19.02	0.14	0	1.904	5852	0.94	0.9
Kepler-371	b	2	34.76	0.2	0	1.8928	5666	0.94	0.99
Kepler-371	c	2	67.97	0.313	0	1.7808	5666	0.94	0.99
Kepler-372	b	3	6.85	0.075	0	1.3552	6509	1.15	1.14
Kepler-372	c	3	20.05	0.154	0	2.0832	6509	1.15	1.14
Kepler-372	d	3	30.09	0.201	0	1.6912	6509	1.15	1.14
Kepler-373	b	2	5.54	0.06	0	1.3552	5787	0.87	0.84
Kepler-373	c	2	16.73	0.126	0	1.2432	5787	0.87	0.84
Kepler-374	b	3	1.9	0.029	0	1.0304	5977	0.84	0.91
Kepler-374	c	3	3.28	0.042	0	1.0976	5977	0.84	0.91
Kepler-374	d	3	5.03	0.056	0	1.3104	5977	0.84	0.91
Kepler-375	b	2	12.13	0.101	0	1.4448	5826	0.89	0.84
Kepler-375	c	2	19.99	0.141	0	2.6432	5826	0.89	0.84
Kepler-376	b	2	4.92	0.057	0	1.064	5900	1.05	1.18
Kepler-376	c	2	14.17	0.115	0	1.792	5900	1.05	1.18
Kepler-377	b	2	12.51	0.109	0	1.3888	5949	0.88	1.22
Kepler-377	c	2	27.01	0.182	0	2.0608	5949	0.88	1.22
Kepler-378	b	2	16.09	0.112	0	0.7504	4661	0.96	0.67
Kepler-378	c	2	28.91	0.166	0	0.6944	4661	0.96	0.67
Kepler-379	b	2	20.1	0.152	0	1.6576	6054	1.08	1.31
Kepler-379	c	2	62.78	0.326	0	2.2848	6054	1.08	1.31
Kepler-38	b	1	105.6	0.4632	122.112	4.3008	5623	0.94	1.75
Kepler-380	b	2	3.93	0.05	0	1.1872	6045	1.05	1.22
Kepler-380	c	2	7.63	0.078	0	1.2656	6045	1.05	1.22
Kepler-381	b	2	5.63	0.066	0	0.9856	6152	1.34	1.57
Kepler-381	c	2	13.39	0.117	0	1.12	6152	1.34	1.57
Kepler-382	b	2	5.26	0.055	0	1.3216	5600	0.8	0.94
Kepler-382	c	2	12.16	0.097	0	1.5904	5600	0.8	0.94

(Continued)

(*Continued*)

Host	Letter	Number	p (days)	d (AU)	M (Earths)	R (Earths)	Temperature (K)	M_s (Suns)	R_s (Suns)
Kepler-383	b	2	12.9	0.095	0	1.3216	4710	0.67	0.67
Kepler-383	c	2	31.2	0.172	0	1.2432	4710	0.67	0.67
Kepler-384	b	2	22.6	0.148	0	1.12	5577	0.76	0.88
Kepler-384	c	2	45.35	0.236	0	1.1312	5577	0.76	0.88
Kepler-385	b	2	10.04	0.097	0	2.7328	6326	1.09	1.13
Kepler-385	c	2	15.16	0.127	0	3.0352	6326	1.09	1.13
Kepler-386	b	2	12.31	0.096	0	1.3888	5178	0.74	0.77
Kepler-386	c	2	25.19	0.155	0	1.5792	5178	0.74	0.77
Kepler-387	b	2	6.79	0.068	0	1.0304	5774	1.03	1.05
Kepler-387	c	2	11.84	0.098	0	0.8848	5774	1.03	1.05
Kepler-388	b	2	3.17	0.036	0	0.8064	4498	0.59	0.59
Kepler-388	c	2	13.3	0.093	0	0.8624	4498	0.59	0.59
Kepler-389	b	2	3.24	0.041	0	1.512	5376	0.78	0.79
Kepler-389	c	2	14.51	0.11	0	1.456	5376	0.78	0.79
Kepler-39	b	1	21.09	0.164	6391.8	13.888	6350	1.29	1.4
Kepler-390	b	2	6.74	0.065	0	0.8176	5166	0.67	0.78
Kepler-390	c	2	13.06	0.101	0	0.784	5166	0.67	0.78
Kepler-391	b	2	7.42	0.082	0	3.192	4940	1.22	3.57
Kepler-391	c	2	20.49	0.161	0	3.5392	4940	1.22	3.57
Kepler-392	b	2	5.34	0.059	0	0.9968	5938	1.13	1.13
Kepler-392	c	2	10.42	0.093	0	1.0976	5938	1.13	1.13
Kepler-393	b	2	9.18	0.091	0	1.288	6189	1.32	1.38
Kepler-393	c	2	14.61	0.124	0	1.3328	6189	1.32	1.38
Kepler-394	b	2	8.01	0.083	0	1.6016	6402	1.11	1.13
Kepler-394	c	2	12.13	0.11	0	1.6576	6402	1.11	1.13
Kepler-395	b	2	7.05	0.061	0	1.0304	4262	0.53	0.56
Kepler-395	c	2	34.99	0.177	0	1.3216	4262	0.53	0.56
Kepler-397	b	2	22.25	0.144	0	2.4528	5307		0.77
Kepler-397	c	2	135.5	0.48	0	6.1712	5307		0.77
Kepler-398	b	3	4.08	0.044	0	0.9296	4493		0.61
Kepler-398	c	3	11.42	0.087	0	1.008	4493		0.61
Kepler-399	b	3	14.43	0.103	0	0.9632	5502		0.68
Kepler-399	c	3	26.68	0.155	0	1.4336	5502		0.68
Kepler-399	d	3	58.03	0.261	0	1.8928	5502		0.68
Kepler-4	b	1	3.21	0.0456	24.486	3.9984	5857	1.22	1.49
Kepler-40	b	1	6.87	0.08	699.6	13.104	6510	1.48	2.13
Kepler-400	b	2	9.02	0.087	0	1.6464	5886		1.15
Kepler-400	c	2	17.34	0.134	0	1.4896	5886		1.15
Kepler-401	b	3	14.38	0.122	0	1.7136	6117		1.33
Kepler-401	c	3	47.32	0.269	0	2.1504	6117		1.33
Kepler-402	b	4	4.03	0.051	0	1.2208	6090		1.26
Kepler-402	c	4	6.12	0.068	0	1.5568	6090		1.26
Kepler-402	d	4	8.92	0.087	0	1.3776	6090		1.26
Kepler-402	e	4	11.24	0.102	0	1.456	6090		1.26
Kepler-403	b	3	7.03	0.076	0	1.2544	6090		1.33
Kepler-403	c	3	54.28	0.297	0	1.7472	6090		1.33

Kepler-404	b	2	11.83	0.102	0	1.2656	5654		0.88
Kepler-404	c	2	14.75	0.118	0	1.7136	5654		0.88
Kepler-405	b	2	10.61	0.095	0	2.0832	5818		0.89
Kepler-405	c	2	29.73	0.188	0	4.6592	5818		0.89
Kepler-41	b	1	1.86	0.03101	178.08	14.448	5750	1.15	1.29
Kepler-410	Ab	1	17.83	0.1226	0	2.8336	6273	1.21	1.35
Kepler-411	b	2	3.01	0.038	0	1.8816	4974	0.83	0.79
Kepler-412	b	1	1.72	0.02897	299.238	15.0192	5750	1.17	1.29
Kepler-413	b	1	66.26	0.3553	67.098	4.3456	4700	0.82	0.78
Kepler-419	b	2	69.75	0.37	795	10.752	6430	1.39	1.74
Kepler-419	c	2	675.47	1.68	2321.4	0	6430	1.39	1.74
Kepler-42	b	3	1.21	0.0116	0	0.784	3068	0.13	0.17
Kepler-42	c	3	0.45	0.006	0	0.728	3068	0.13	0.17
Kepler-42	d	3	1.87	0.0154	0	0.5712	3068	0.13	0.17
Kepler-420	b	1	86.65	0.382	461.1	10.528	5520	0.99	1.13
Kepler-421	b	1	704.2	1.219	0	4.1552	5308	0.79	0.76
Kepler-422	b	1	7.89	0.082	136.74	12.88	5972	1.15	1.24
Kepler-423	b	1	2.68	0.03585	189.21	13.3504	5560	0.85	0.95
Kepler-424	b	2	3.31	0.044	327.54	9.968	5460	1.01	0.94
Kepler-424	c	2	223.3	0.73	2216.46	0	5460	1.01	0.94
Kepler-425	b	1	3.8	0.0464	79.5	10.9536	5170	0.93	0.86
Kepler-426	b	1	3.22	0.0414	108.12	12.208	5725	0.91	0.92
Kepler-427	b	1	10.29	0.091	92.22	13.776	5800	0.96	1.35
Kepler-428	b	1	3.53	0.0433	403.86	12.096	5150	0.87	0.8
Kepler-43	b	1	3.02	0.046	1027.14	13.6528	6041	1.32	1.42
Kepler-430	b	2	35.97	0.2244	0	3.248	5884	1.17	1.49
Kepler-430	c	2	110.98	0.4757	0	1.792	5884	1.17	1.49
Kepler-431	b	3	6.8	0.0719	0	0.7616	6004	1.07	1.09
Kepler-431	c	3	8.7	0.0847	0	0.672	6004	1.07	1.09
Kepler-431	d	3	11.92	0.1045	0	1.12	6004	1.07	1.09
Kepler-432	b	2	52.5	0.301	1720.38	12.824	4995	1.32	4.06
Kepler-433	b	1	5.33	0.0679	896.76	16.24	6360	1.46	2.26
Kepler-434	b	1	12.87	0.1143	909.48	12.656	5977	1.2	1.38
Kepler-435	b	1	8.6	0.0948	267.12	22.288	6161	1.54	3.21
Kepler-436	b	2	64	0.339	0	2.688	4651	0.73	0.7
Kepler-437	b	1	66.65	0.288	0	2.128	4551	0.71	0.68
Kepler-438	b	1	35.23	0.166	0	1.12	3748	0.54	0.52
Kepler-439	b	1	178.14	0.563	0	2.24	5431	0.88	0.87
Kepler-44	b	1	3.25	0.0446	318	12.208	5800	1.12	1.35
Kepler-440	b	1	101.11	0.242	0	1.904	4134	0.57	0.56
Kepler-441	b	1	207.25	0.64	0	1.68	4340	0.57	0.55
Kepler-442	b	1	112.31	0.409	0	1.344	4402	0.61	0.6
Kepler-443	b	1	177.67	0.495	0	2.352	4723	0.74	0.71
Kepler-444	b	5	3.6	0.04178	0	0.4032	5046	0.76	0.75
Kepler-444	c	5	4.55	0.04881	0	0.4928	5046	0.76	0.75
Kepler-444	d	5	6.19	0.06	0	0.5264	5046	0.76	0.75
Kepler-444	e	5	7.74	0.0696	0	0.5488	5046	0.76	0.75
Kepler-444	f	5	9.74	0.0811	0	0.7392	5046	0.76	0.75
Kepler-447	b	1	7.79	0.0769	435.66	18.48	5493	1	1.05
Kepler-45	b	1	2.46	0.03	160.59	10.752	3820	0.59	0.55

(*Continued*)

A-17

(*Continued*)

Host	Letter	Number	p (days)	d (AU)	M (Earths)	R (Earths)	Temperature (K)	M_s (Suns)	R_s (Suns)
Kepler-452	b	1	384.84	1.046	0	1.624	5757	1.04	1.11
Kepler-453	b	1	240.5	0.7903	0.20034	6.1936	5527	0.94	0.83
Kepler-454	b	2	10.57	0.0954	6.84336	2.3632	5701	1.03	1.07
Kepler-46	b	3	33.6	0.1968	1908	9.0496	5309	0.9	0.79
Kepler-46	c	3	57.01	0.2799	119.568	0	5309	0.9	0.79
Kepler-46	d	3	6.77	0.067	0	1.6352	5309	0.9	0.79
Kepler-47	b	2	49.53	0.2962	636	3.024	5636	1.05	0.96
Kepler-47	c	2	303.14	0.991	8904	4.6032	5636	1.05	0.96
Kepler-49	d	4	2.58	0.031	0	1.6016	4252	0.55	0.56
Kepler-49	e	4	18.6	0.116	0	1.5568	4252	0.55	0.56
Kepler-5	b	1	3.55	0.0538	671.298	15.9712	6297	1.37	1.79
Kepler-50	b	2	7.81	0.077	0	1.7136	6225	1.24	1.58
Kepler-50	c	2	9.38	0.087	0	2.1728	6225	1.24	1.58
Kepler-51	b	3	45.15	0.2514	2.226	7.0896	6018	1.04	0.94
Kepler-51	c	3	85.31	0.384	4.134	8.9936	6018	1.04	0.94
Kepler-51	d	3	130.19	0.509	7.632	9.688	6018	1.04	0.94
Kepler-52	d	3	36.45	0.182	0	1.9488	4263	0.54	0.56
Kepler-53	d	3	9.75	0.091	0	2.1168	5858	0.98	0.89
Kepler-539	b	2	125.63	0.4988	308.46	8.3664	5820	1.05	0.95
Kepler-539	c	2	1000	2.42	763.2	0	5820	1.05	0.95
Kepler-54	d	3	21	0.126	0	1.5232	4252	0.51	0.55
Kepler-55	d	5	2.21	0.029	0	1.5904	4503	0.62	0.62
Kepler-55	e	5	4.62	0.048	0	1.5456	4503	0.62	0.62
Kepler-55	f	5	10.2	0.081	0	1.5904	4503	0.62	0.62
Kepler-56	b	2	10.5	0.1028	22.26	6.5072	4840	1.32	4.23
Kepler-56	c	2	21.4	0.1652	181.26	9.7888	4840	1.32	4.23
Kepler-58	d	3	40.1	0.236	0	2.9344	6099	0.95	1.13
Kepler-6	b	1	3.23	0.04852	212.424	14.6048	5647	1.21	1.39
Kepler-62	b	5	5.71	0.0553	9.54	1.3104	4925	0.69	0.64
Kepler-62	c	5	12.44	0.0929	4.134	0.5376	4925	0.69	0.64
Kepler-62	d	5	18.16	0.12	13.992	1.9488	4925	0.69	0.64
Kepler-62	e	5	122.39	0.427	35.934	1.6128	4925	0.69	0.64
Kepler-62	f	5	267.29	0.718	34.98	1.4112	4925	0.69	0.64
Kepler-63	b	1	9.43	0.08	120.204	6.104	5576	0.98	0.9
Kepler-65	b	3	2.15	0.035	0	1.4224	6211	1.25	1.41
Kepler-65	c	3	5.86	0.068	0	2.576	6211	1.25	1.41
Kepler-65	d	3	8.13	0.084	0	1.5232	6211	1.25	1.41
Kepler-66	b	1	17.82	0.1352	0	2.8	5962	1.04	0.97
Kepler-67	b	1	15.73	0.1171	0	2.9344	5331	0.86	0.78
Kepler-68	b	3	5.4	0.0617	8.268	2.3072	5793	1.08	1.24
Kepler-68	c	3	9.61	0.09059	4.77	0.952	5793	1.08	1.24
Kepler-68	d	3	580	1.4	301.146	0	5793	1.08	1.24
Kepler-69	b	2	13.72	0.094	0	2.24	5638	0.81	0.93
Kepler-69	c	2	242.46	0.64	0	1.7136	5638	0.81	0.93
Kepler-7	b	1	4.89	0.06067	140.238	18.1664	5933	1.36	1.97
Kepler-74	b	1	7.34	0.0781	200.34	10.752	6000	1.18	1.12
Kepler-75	b	1	8.88	0.0818	3211.8	11.76	5200	0.91	0.89
Kepler-76	b	1	1.54	0.0274	639.18	15.232	6409	1.2	1.32

Kepler-77	b	1	3.58	0.04501	136.74	10.752	5520	0.95	0.99
Kepler-79	b	4	13.48	0.117	10.9074	3.472	6174	1.17	1.3
Kepler-79	c	4	27.4	0.187	6.042	3.7184	6174	1.17	1.3
Kepler-79	d	4	52.09	0.287	6.042	7.1568	6174	1.17	1.3
Kepler-79	e	4	81.07	0.386	4.134	3.4832	6174	1.17	1.3
Kepler-8	b	1	3.52	0.0474	187.62	15.8592	6213	1.21	1.49
Kepler-80	d	5	3.07	0.037	0	1.736	4613	0.72	0.64
Kepler-80	e	5	4.65	0.049	0	1.6352	4613	0.72	0.64
Kepler-81	d	3	20.84	0.128	0	1.2096	4500	0.64	0.59
Kepler-82	d	4	2.38	0.034	0	1.7696	5428	0.85	0.94
Kepler-82	e	4	5.9	0.063	0	2.464	5428	0.85	0.94
Kepler-83	d	3	5.17	0.051	0	1.9376	4648	0.66	0.59
Kepler-84	d	5	4.22	0.052	0	1.3776	6031	1	1.17
Kepler-84	e	5	27.43	0.181	0	2.5984	6031	1	1.17
Kepler-84	f	5	44.55	0.25	0	2.1952	6031	1	1.17
Kepler-85	d	4	17.91	0.13	0	1.1984	5436	0.92	0.89
Kepler-85	e	4	25.22	0.163	0	1.2656	5436	0.92	0.89
Kepler-87	b	2	114.74	0.481	324.36	13.4848	5600	1.1	1.82
Kepler-87	c	2	191.23	0.676	6.36	6.1376	5600	1.1	1.82
Kepler-9	b	3	19.24	0.14	80.136	9.4304	5777	1.07	1.02
Kepler-9	c	3	38.91	0.225	54.378	9.2176	5777	1.07	1.02
Kepler-9	d	3	1.59	0.0273	0	1.6352	5777	1.07	1.02
Kepler-91	b	1	6.25	0.0731	257.58	15.3104	4550	1.31	6.3
Kepler-93	b	2	4.73	0.053	4.0227	1.4784	5669	0.91	0.92

Appendix B

Finding Possibly Habitable Planets

Since its launch in 2009, The *Kepler* spacecraft has discovered over 2000 *exoplanets*, planets orbiting stars other than our Sun. *Kepler* uses the *Transit-Method* for exoplanet detection. As shown in Figure B.1 below, a sensitive light—detector called a photometer on the spacecraft measures small changes in a star's brightness occurring when a planet passes in front of it. The measured amount of change in brightness and time it lasts allows the size of the planet and the period of its orbit around the star to be determined which also allows the orbital distance to be calculated.

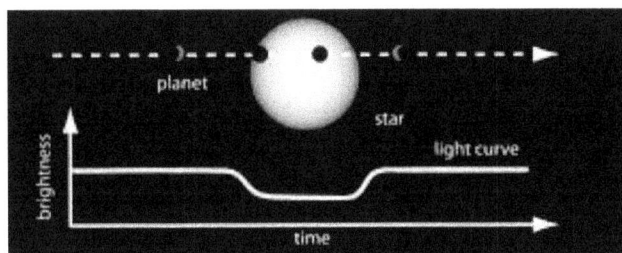

Figure B.1. The *Transit-Method* for exoplanet detection.

The *Host Name*, identifies the planetary system; the *Planet Letter*, the star is-a, the first planet-b, the second-c and so on; the *Number of Planets* in each system; the *Orbital Period* in Earth-Days; the *Orbital Distance* in AU; the *Mass*; and *Radius* of each planet compared to Earth = 1; the Kelvin *Temperature* of the Star and the *Mass*; and *Radius* of each Star compared to the Sun = 1.

1	2	3	4	5	6	7	8	9	10
Host Name	Planet Letter	Number of Planets	Orbital Period (days)	Orbital Distance (AU)	Planet Mass (E = 1)	Planet Radius (E = 1)	Star Temp (K)	Star Mass (S = 1)	Star Rad. (S = 1)

doi:10.1088/2514-3433/ad7c3ech8
B-1

First, sort your data in order of increasing Planet Radius (column 7 of your data) and determine how many of each planet type (listed Table B.1 below) you have in your data set. Record in Table B.1 how many of each planet type you find. Then calculate the % of the total number of planets for each planet type. You do this by dividing the number of planets of one type by the total number of all planets examined then multiplying by 100%

You may now rule out any planets in the data with a radius > 2.3 times that of Earth. This eliminates all planets except *Earths* and *Super-Earths* with a 15% margin for error above the upper limit for *Super-Earths*. If there is no radius given for a planet in your latter, compare its mass (column 6) to Earth's to determine whether or not to include it.

Planets large enough to be *ice-giants* or *gas-giants* like Neptune or Jupiter are not going to be able to support life like that found on our Earth. To be *Possibly Habitable* a planet must not only be *in its star's habitable zone*, but it also must be an *Earth-like* planet or possibly a *Super-Earth* with a rock/metal composition and thus a solid surface.

Next, sort your remaining data in order of increasing Star Temperature (column 8) so you can identify Spectral Class of each star, either M, K, G or F, as given in Table B.2.

Table B.1. Planet Types.

Planet Type	Radius (Earth = 1)	Number	% of Total = ___
Earth-like	< 1.25		
Super-Earth	1.25-2		
Neptune-Like (ice giant)	2–6		
Jupiter-like (gas giant)	6–15		
Larger	> 15		

Table B.2. Star Types.

Spectral Class	Star Type	Temp. (K)	Mass (Sun = 1)
M	Red Dwarf	< 4000	up to about 0.5
K	Orange	4000–5300	about 0.45–0.8
G	Yellow (Sun-like)	> 5300–6000	0.8–1.2
F	Green/White	> 6000	About 1 to 1.4

Table B.3. Habitable Zones For Different Spectral Class (Star Types).

Spectral Class	Approximate Habitable Zone (AU)
M (Red Dwarf)	0.1–0.5
K (Orange)	0.2–1.2
G (Yellow-Sun-like)	0.4–2.5
F (Green/White)	0.7–4

Table B.4. *Possibly Habitable* Planets.

Host Name	Planet letter	Orbital Distance (AU)	Planet Radius (Earth = 1)	Spectral Class	Habitable Zone (AU)

Now it's time to go hunting for planets in habitable zones! The *habitable zone* is the approximate range of distances from a star in which water can exist as a liquid. Table B.3 below gives the approximate habitable zone for stars of each Spectral Class.

Separate your data set into four (4) different sets, one for each spectral class. Then sort the data for each spectral class in order of increasing orbital distance (column 5). Look at the planets in each of your (4) data sets to see if any have an orbital distance (column 5) within the habitable zone *for its spectral class*. Now, enter the information for each possibly habitable planet you find in Table B.4.

Questions

1. How many *Possibly Habitable* planets did you find?
2. About what percentage of the planets you analyzed were determined to be *Possibly Habitable*? Divide the number of *Possibly Habitable* planets by the *total* number of planets you analyzed then multiply by 100%
3. Which type of star has the most *Possibly Habitable* planets?
4. Which type of star has the least *Possibly Habitable* planets?
5. Now that the *Kepler* Mission has found *Possibly Habitable* planets, future missions are planned to attempt to examine the spectra of these planets. What chemical-signatures, if detected, could indicated that there may indeed be life on these planets?

1. Which is NOT a condition for a planet to be considered *Possibly Habitable*?
   A. Orbiting within its star's habitable zone
   B. Being similar in size to Earth or a Super Earth
   C. Orbiting a Sun-like star.
   D. [All of the above are conditions]

2. Based on the *Kepler* data, which **Planet Type** is most abundant?
   A. Earth-like
   B. Super-Earth
   C. Neptune-like (ice-giant)
   D. Jupiter-like (gas-giant)
   E. Larger

3. Which type of star has the largest habitable zone?
   A. M
   B. K
   C. G
   D. F

4. According to your class results, which type of star DOES NOT appear to have any *Possibly Habitable* planets in orbit?
   A. M
   B. K
   C. G
   D. F

5. According to your class results, which type of star has the most *Possibly Habitable* planets in orbit?
   A. M
   B. K
   C. G
   D. F

Student Exoplanet Projects Using Data from the Kepler Mission

Michael C LoPresto

Appendix C

Planetary Temperatures Radiation, Albedo and the Greenhouse Effect

In this experiment you will explore the effects on the temperature of a planet of radiation from the Sun, the albedo or reflectivity of a planet's surface and the greenhouse effect. You will calculate the effects that each of these processes have on a planet's temperature then compare your calculations to actual measured temperatures of the planets.

C.1 Radiation

The main source of energy in the solar system is *radiation* from the Sun. This is the main factor that affects the temperature of a planet. The *Stefan–Boltzmann Radiation Law*, states that the rate at which an object emits radiation through its surface area is proportional to its temperature to the fourth power, $J \sim T^4$. As the Sun emits radiation out into the solar system, a planet will absorb just a small fraction of the radiation that has traveled out to its distance that depends on the planet's size. An expression that can be derived for the temperature of a planet a distance, d, from the Sun based on the Stefan–Boltzmann Law, the Sun's temperature of 5800 K, the Sun's Radius of 6.95×10^8 m is

$$T_r = 279/(d)^{0.5} \tag{C.1}$$

Use the **orbital radius** (a planet's average distance from the Sun) of each planet as, d, found in Table C.1, to calculate a theoretical temperature for each planet based on the radiation it absorbs from the Sun and record them in the column labeled **Theoretical Temperature** in Table C.1.

Raising a number to the 0.5 (one-half) power is the same as taking the square root, so the calculation amounts to 279 divided by the square root of the planet's orbital radius. You may round off any decimals in your calculated temperatures to the nearest whole number.

doi:10.1088/2514-3433/ad7c3ech9   C-1   © IOP Publishing Ltd 2024. All rights,
including for text and data mining (TDM), artificial intelligence (AI) training, and similar technologies, are reserved.

Table C.1. Planetary Temperatures from Radiation.

Object	Orbital Radius d (AU)	Theoretical (Calculated) Temperature (from r = radiation) $T_r = 279(d)^{0.5}$ Tr (K)	Observed (Measured) Temperature To To (K)	% Error %
Mercury	0.39		452	
Venus	0.72		726	
Earth	1.00		285	
Mars	1.52		230	
Jupiter	5.2		120	
Saturn	9.5		88	
Uranus	19.2		59	
Neptune	30.1		48	
Pluto	39.5		37	

C.2 Percentage Error

Your temperatures are *theoretical* and must now be compared to measured or *observed* experimental temperatures for each planet found in the column labeled **Observed Temperature** in Table C.1. Now, for each planet, calculate the percentage error that your theoretical temperature, T, is off from the observed temperature, T_o, with the formula;

$$([T - T_o]/T_o) \times 100\% \tag{C.2}$$

This is simply the difference between your theoretical and observed temperatures divided by the actual observed temperature. Note: Whether the number in the parenthesis $[T - T_o]$ comes out positive or negative, just use it as positive in your calculation. We are mostly concerned with how far off the theoretical temperatures are from the observed temperatures, not whether they are too high or two low. Enter your % errors in the last column of Table C.1.

Now examine the % errors you calculated. Any theoretical temperature that is <5% away from the experimental temperature can be considered a satisfactory value. This means that the planet's temperature can be considered mostly due to radiation from the Sun. If the error is more than 5%, then we must consider other factors that affect a planet's temperature. Circle any errors that you find that are greater than 5%.

C.3 Albedo

The *albedo*, a, of a planet is the percentage of the radiation incident upon the planet that the planet reflects back out into space rather than absorbs. Earth's albedo is $a = 0.31$, meaning that it reflects away 31% of the radiation that comes to it from the

Table C.2. Planetary Temperatures from Radiation and Albedo.

Object	Albedo A	Theoretical (Calculated) Temperature (From a=albedo and r = radiation) $T_a = T_r \times (1 - a)^{0.25}$ T_a (K)	% Error %
Mercury	0.12		
Venus	0.75		
Earth	0.31		
Mars	0.25		
Jupiter	0.34		
Saturn	0.34		
Uranus	0.3		
Neptune	0.3		
Pluto	0.4		

Sun. The albedo for each planet is found in the column labeled **Albedo** in Table C.2. Now, ONLY FOR PLANETS THAT HAD > 5% ERROR WHEN THEIR TEMPERATURE WAS CALCULATED BY RADIATION, use the following expression to calculate a new theoretical temperature for a planet;

$$T_a = T_r \times (1 - a)^{0.25} \tag{C.3}$$

First subtract the albedo, a, from 1 then raise that number to the 0.25 (one-fourth) power. This is the same as taking the square root *twice*. Then multiply your result times the theoretical temperatures you already have for radiation alone, T_r from Table C.1. This will be theoretical temperature of a planet based on the effects of radiation *and* albedo. Enter your new **Theoretical Temperatures** (from albedo and radiation) in their column in Table C.2. Again, do this ONLY for planets that had greater than 5% error in their temperature calculated from radiation only.

Now, for each planet for which you calculated the effect of albedo on its temperature, use Equation (C.2) to calculate the percentage error that your new theoretical temperature, T_a, in Table C.2 is off from the *original observed temperature* in Table C.1 and enter your new % errors in their column next to (to the right of) the theoretical temperatures from radiation *and* albedo in Table C.2. Once again, any error 5% or less is acceptable, but any error much more than that must still be corrected for other factors.

C.4 The Greenhouse Effect

After a planet absorbs radiation from the Sun, it reemits it. Certain gases in a planet's atmosphere, such as water vapor and carbon dioxide absorb some of this reemitted energy before the rest of it escapes back out into space. This is called the

Table C.3. Planetary Temperatures from Radiation, Albedo and the Greenhouse Effect.

Object	τ	Theoretical (Calculated) Temperature (From $a =$ Albedo, $r =$ Radiation and $g =$ Greenhouse Effect) $T_g = T_a \times (1 + \tau)^{0.25}$	% Error %
Venus	92		
Earth	1		
Mars	0.2		

greenhouse effect. The greenhouse effect can be considered in our calculations by using the following formula:

$$T_g = T_a \times (1 + \tau)^{0.25} \qquad (C.4)$$

The *thickness* of greenhouse gases, τ in Equation (C.4), is a comparison between the amounts of greenhouse gases in a planet's atmosphere to those on Earth (where $\tau = 1$). In Mars' very thin atmosphere, $\tau = 0.2$ but in Venus' very thick atmosphere $\tau = 92$. Use Equation (C.4) to include the greenhouse effect in your calculation of the temperature of any planet that still had >5 % error in its theoretical temperature after considering albedo in Table C.2. Enter your results in the column labeled **Theoretical Temperatures** (from albedo, radiation and greenhouse effect) in Table C.3.

This time *add* the thickness, τ, to 1 then raise that number to the 0.25 (one-fourth) power by taking the square root *twice*. Then finally, multiply your result times the theoretical temperature you already have for radiation *and* albedo from Table C.2. Also, use Equation (C.2) to calculate a new % error, how far T_g is off from the original T_o in Table C.1, and enter it in the last column of Table C.3.

C.5 Questions

1 Which planetary temperatures could be satisfactorily calculated by considering radiation alone?

2 Which planetary temperatures were satisfactorily calculated by including the effect of albedo in addition to radiation?

How does albedo affect a planets' temperature? To answer this, look at the change in the theoretical temperatures of the planets for which you included the effect of albedo.

Why does albedo have this effect on a planet's temperature?

Why do you think Venus has such a high albedo? Why do you think Pluto does?

3 Which planetary temperatures were only calculated satisfactorily by including the greenhouse effect in addition to radiation and albedo?

What affect does the greenhouse effect have on a planet's temperature?

Why does the greenhouse effect affect a planet's temperature this way?

How does the greenhouse effect on Earth compare to that on Mars? Why is this the case?

How does the greenhouse effect on Earth compare to that on Venus? Why is this the case?

1. Which can have an effect on the temperature of *any* planet?
   A. radiation;
   B. albedo;
   C. the greenhouse effect;
   D. A and B only;
   E. All of the above.

2. Albedo is the percentage of the energy from the Sun that
   A. is absorbed by a planet's surface;
   B. is reflected by a planet's surface;
   C. is absorbed by a planet's atmosphere.

3. The greenhouse effect occurs
   A. in the atmospheres of all planets;
   B. in the atmospheres of terrestrial planets only;
   C. in the atmosphere of Jovian planets only.

4. Which planet experiences very little greenhouse effect?
   A. Earth;
   B. Venus;
   C. Mars.

5. Which planet experiences a tremendous amount of greenhouse effect?
   A. Earth;
   B. Venus;
   C. Mars.

Appendix D

Kepler Density Spreadsheet Data

This appendix is also available as a spreadsheet from [doi link 10.1088/978-0-7503-6290-0]

pl_name	pl_discmethod	pl_orbsmax	pl_bmassj	pl_radj
Kepler-435 b	Transit	0.0948	0.84	1.99
Kepler-12 b	Transit	0.0553	0.432	1.754
Kepler-447 b	Transit	0.0769	1.37	1.65
Kepler-7 b	Transit	0.06067	0.441	1.622
Kepler-433 b	Transit	0.0679	2.82	1.45
Kepler-5 b	Transit	0.0538	2.111	1.426
Kepler-8 b	Transit	0.0474	0.59	1.416
Kepler-91 b	Transit	0.0731	0.81	1.367
Kepler-76 b	Orbital Brightness Modulation	0.0274	2.01	1.36
Kepler-412 b	Transit	0.02897	0.941	1.341
Kepler-17 b	Transit	0.02591	2.45	1.31
Kepler-6 b	Transit	0.04852	0.668	1.304
Kepler-41 b	Transit	0.03101	0.56	1.29
Kepler-39 b	Transit	0.164	20.1	1.24
Kepler-427 b	Transit	0.091	0.29	1.23
Kepler-43 b	Transit	0.046	3.23	1.219
Kepler-87 b	Transit	0.481	1.02	1.204
Kepler-418 b	Transit		1.1	1.2
Kepler-423 b	Transit	0.03585	0.595	1.192
Kepler-40 b	Transit	0.08	2.2	1.17
Kepler-422 b	Transit	0.082	0.43	1.15
Kepler-432 b	Transit	0.301	5.41	1.145
Kepler-14 b	Transit		8.4	1.136
Kepler-434 b	Transit	0.1143	2.86	1.13
Kepler-117 c	Transit	0.2804	1.84	1.101

(*Continued*)

(*Continued*)

pl_name	pl_discmethod	pl_orbsmax	pl_bmassj	pl_radj
Kepler-30 c	Transit	0.3	2.01	1.097
Kepler-44 b	Transit	0.0446	1	1.09
Kepler-426 b	Transit	0.0414	0.34	1.09
Kepler-428 b	Transit	0.0433	1.27	1.08
Kepler-1658 b	Transit	0.0544	5.88	1.07
Kepler-1647 b	Transit	2.7205	1.51968	1.059
Kepler-75 b	Transit	0.0818	10.1	1.05
Kepler-289 c	Transit	0.51	0.42	1.034
Kepler-425 b	Transit	0.0464	0.25	0.978
Kepler-419 b	Transit	0.37	2.5	0.96
Kepler-15 b	Transit	0.05714	0.66	0.96
Kepler-74 b	Transit	0.0781	0.63	0.96
Kepler-45 b	Transit	0.03	0.505	0.96
Kepler-77 b	Transit	0.04501	0.43	0.96
Kepler-424 b	Transit	0.044	1.03	0.89
Kepler-56 c	Transit	0.1652	0.57	0.874
Kepler-51 d	Transit	0.509	0.024	0.865
Kepler-1654 b	Transit	2.026	0.5	0.819
Kepler-46 b	Transit	0.1968	6	0.808
Kepler-51 c	Transit	0.384	0.013	0.803
Kepler-30 d	Transit	0.5	0.073	0.785
Kepler-34 b	Transit	1.0896	0.22	0.764
Kepler-16 b	Transit	0.7048	0.333	0.754
Kepler-539 b	Transit	0.4988	0.97	0.747
Kepler-9 b	Transit	0.143	0.13655	0.74
Kepler-35 b	Transit	0.60347	0.127	0.728
Kepler-9 c	Transit	0.227	0.09408	0.721
Kepler-117 b	Transit	0.1445	0.094	0.719
Kepler-79 d	Transit	0.287	·0.019	0.639
Kepler-177 c	Transit		0.024	0.633
Kepler-51 b	Transit	0.2514	0.007	0.633
Kepler-47 d	Transit	0.6992	0.05984	0.628
Kepler-18 d	Transit	0.1172	0.052	0.623
Kepler-56 b	Transit	0.1028	0.07	0.581
Kepler-453 b	Transit	0.7903	0.05034	0.553
Kepler-87 c	Transit	0.676	0.02	0.548
Kepler-63 b	Transit	0.08	0.378	0.545
Kepler-101 b	Transit	0.0474	0.16	0.51
Kepler-238 e	Transit		0.534	0.5
Kepler-18 c	Transit	0.0752	0.054	0.49
Kepler-328 c	Transit		0.124	0.482
Kepler-82 c	Transit	0.2626	0.04373	0.476
Kepler-31 c	Transit	0.26	4.7	0.473
Kepler-396 c	Transit		0.056	0.473

Kepler-223 d	Transit		0.02517	0.467
Kepler-25 c	Transit		0.0479	0.465
Kepler-103 c	Transit		0.114	0.459
Kepler-1656 b	Transit	0.197	0.15291	0.448
Kepler-27 c	Transit	0.191	13.8	0.437
Kepler-47 c	Transit	0.9638	0.00997	0.415
Kepler-223 e	Transit		0.0151	0.41
Kepler-411 c	Transit	0.0739	0.08306	0.394
Kepler-413 b	Transit	0.3553	0.211	0.388
Kepler-145 c	Transit		0.25	0.385
Kepler-38 b	Transit	0.4632	0.384	0.384
Kepler-279 c	Transit		0.155	0.384
Kepler-11 e	Transit	0.195	0.025	0.374
Kepler-82 b	Transit	0.1683	0.03823	0.363
Kepler-27 b	Transit	0.118	9.11	0.357
Kepler-4 b	Transit	0.0456	0.077	0.357
Kepler-31 d	Transit	0.4	6.8	0.348
Kepler-30 b	Transit	0.18	0.036	0.348
Kepler-79 c	Transit	0.187	0.019	0.332
Kepler-36 c	Transit	0.1283	0.025	0.328
Kepler-28 b	Transit	0.062	1.51	0.321
Kepler-305 b	Transit		0.033	0.321
Kepler-92 b	Transit		0.202	0.313
Kepler-94 b	Transit		0.034	0.313
Kepler-396 b	Transit		0.238	0.312
Kepler-79 e	Transit	0.386	0.013	0.311
Kepler-79 b	Transit	0.117	0.0343	0.31
Kepler-223 c	Transit		0.01605	0.307
Kepler-95 b	Transit		0.041	0.305
Kepler-28 c	Transit	0.081	1.36	0.303
Kepler-103 b	Transit		0.031	0.301
Kepler-277 c	Transit		0.202	0.3
Kepler-29 b	Transit		0.01419	0.299
Kepler-11 g	Transit	0.466	0.079	0.297
Kepler-411 d	Transit	0.279	0.04782	0.296
Kepler-305 c	Transit		0.019	0.294
Kepler-23 c	Transit	0.099	2.7	0.285
Kepler-53 c	Transit		15.74	0.283
Kepler-29 c	Transit		0.01259	0.28
Kepler-11 d	Transit	0.155	0.023	0.278
Kepler-282 e	Transit		0.177	0.277
Kepler-279 d	Transit		0.118	0.277
Kepler-350 c	Transit		0.019	0.277
Kepler-20 c	Transit	0.0949	0.04012	0.272
Kepler-47 b	Transit	0.2877	0.00651	0.272

(*Continued*)

(Continued)

pl_name	pl_discmethod	pl_orbsmax	pl_bmassj	pl_radj
Kepler-414 c	Transit Timing Variations		0.094	0.269
Kepler-223 b	Transit		0.02328	0.267
Kepler-277 b	Transit		0.275	0.261
Kepler-276 c	Transit		0.052	0.259
Kepler-107 e	Transit	0.12639	0.02706	0.259
Kepler-177 b	Transit		0.006	0.259
Kepler-53 b	Transit		18.41	0.258
Kepler-11 c	Transit	0.107	0.009	0.256
Kepler-58 c	Transit		2.19	0.255
Kepler-24 c	Transit	0.106	1.6	0.25
Kepler-276 d	Transit		0.051	0.25
Kepler-350 d	Transit		0.047	0.25
Kepler-58 b	Transit		1.39	0.248
Kepler-26 b	Transit		0.01611	0.248
Kepler-20 d	Transit	0.3506	0.03168	0.245
Kepler-25 b	Transit		0.0275	0.245
Kepler-80 c	Transit	0.0792	0.02121	0.244
Kepler-49 b	Transit		0.98	0.243
Kepler-26 c	Transit		0.01951	0.243
Kepler-48 c	Transit		0.04597	0.242
Kepler-289 d	Transit	0.33	0.013	0.239
Kepler-96 b	Transit		0.027	0.238
Kepler-80 b	Transit	0.0648	0.0218	0.238
Kepler-145 b	Transit		0.117	0.236
Kepler-92 c	Transit		0.019	0.232
Kepler-106 e	Transit		0.035	0.228
Kepler-49 c	Transit		0.72	0.227
Kepler-114 d	Transit		0.012	0.226
Kepler-109 c	Transit		0.06859	0.225
Kepler-106 c	Transit		0.033	0.223
Kepler-11 f	Transit	0.25	0.006	0.222
Kepler-416 b	Transit Timing Variations		0.183	0.221
Kepler-282 d	Transit		0.192	0.219
Kepler-55 b	Transit		1.49	0.217
Kepler-307 b	Transit		0.02341	0.217
Kepler-131 b	Transit		0.051	0.215
Kepler-24 b	Transit	0.08	1.6	0.214
Kepler-411 b	Transit	0.0375	0.08055	0.214
Kepler-22 b	Transit	0.849	0.113	0.212
Kepler-109 b	Transit		0.02297	0.211
Kepler-10 c	Transit		0.054	0.21
Kepler-454 b	Transit		0.02	0.21
Kepler-417 b	Transit Timing Variations		0.035	0.206

Kepler-328 b	Transit		0.09	0.205
Kepler-538 b	Transit	0.4669	0.03335	0.198
Kepler-102 e	Transit		0.028	0.198
Kepler-55 c	Transit		1.11	0.197
Kepler-1655 b	Transit	0.103	0.01573	0.197
Kepler-32 b	Transit	0.05	4.1	0.196
Kepler-100 c	Transit		0.02218	0.196
Kepler-307 c	Transit		0.01145	0.196
Kepler-57 b	Transit		18.86	0.195
Kepler-113 c	Transit		0.02737	0.194
Kepler-289 b	Transit	0.21	0.023	0.192
Kepler-52 b	Transit		8.7	0.187
Kepler-54 b	Transit		0.92	0.187
Kepler-48 d	Transit		0.025	0.182
Kepler-32 c	Transit	0.09	0.5	0.178
Kepler-238 f	Transit		0.042	0.178
Kepler-18 b	Transit	0.0447	0.022	0.178
Kepler-60 d	Transit		0.01309	0.178
Kepler-98 b	Transit		0.011	0.178
Kepler-59 c	Transit		1.37	0.177
Kepler-62 d	Transit	0.12	0.044	0.174
Kepler-37 d	Transit		0.03839	0.173
Kepler-23 b	Transit	0.075	0.8	0.17
Kepler-60 c	Transit		0.01211	0.17
Kepler-48 b	Transit		0.0124	0.168
Kepler-20 b	Transit	0.0463	0.03052	0.167
Kepler-52 c	Transit		10.41	0.164
Kepler-113 b	Transit		0.037	0.162
Kepler-11 b	Transit	0.091	0.006	0.161
Kepler-122 f	Transit Timing Variations		0.113	0.156
Kepler-60 b	Transit		0.01318	0.153
Kepler-414 b	Transit Timing Variations		0.011	0.153
Kepler-21 b	Transit	0.042717	0.01598	0.146
Kepler-62 e	Transit	0.427	0.113	0.144
Kepler-100 d	Transit		0.00944	0.144
Kepler-80 e	Transit	0.0491	0.01299	0.143
Kepler-114 c	Transit		0.009	0.143
Kepler-107 c	Transit	0.06064	0.02954	0.142
Kepler-93 b	Transit		0.01	0.14
Kepler-338 e	Transit Timing Variations		0.027	0.139
Kepler-57 c	Transit		6.95	0.138
Kepler-107 b	Transit	0.04544	0.01104	0.137
Kepler-80 d	Transit	0.0372	0.02124	0.136
Kepler-36 b	Transit	0.1153	0.014	0.133

(*Continued*)

(*Continued*)

pl_name	pl_discmethod	pl_orbsmax	pl_bmassj	pl_radj
Kepler-99 b	Transit		0.019	0.132
Kepler-10 b	Transit	0.0172	0.0145	0.132
Kepler-97 b	Transit		0.011	0.132
Kepler-406 b	Transit		0.02	0.128
Kepler-128 b	Transit		0.00242	0.128
Kepler-62 f	Transit	0.718	0.11	0.126
Kepler-128 c	Transit		0.00283	0.12
Kepler-100 b	Transit		0.023	0.118
Kepler-62 b	Transit	0.0553	0.03	0.117
Kepler-105 c	Transit		0.01447	0.117
Kepler-101 c	Transit	0.0684	0.01	0.112
Kepler-54 c	Transit		0.37	0.11
Kepler-415 b	Transit Timing Variations		0.377	0.108
Kepler-138 d	Transit		0.00201	0.108
Kepler-138 c	Transit		0.0062	0.107
Kepler-409 b	Transit		0.06922	0.106
Kepler-102 d	Transit		0.012	0.105
Kepler-78 b	Transit		0.01	0.1
Kepler-59 b	Transit		2.05	0.098
Kepler-407 b	Transit		0.01007	0.095
Kepler-106 d	Transit		0.02549	0.085
Kepler-102 f	Transit		0.01636	0.079
Kepler-107 d	Transit	0.08377	0.01196	0.077
Kepler-406 c	Transit		0.009	0.076
Kepler-131 c	Transit		0.026	0.075
Kepler-106 b	Transit		0.01668	0.073
Kepler-408 b	Transit		0.01573	0.073
Kepler-37 c	Transit		0.03776	0.067
Kepler-102 c	Transit		0.00944	0.052
Kepler-62 c	Transit	0.0929	0.013	0.048
Kepler-138 b	Transit		0.00021	0.047
Kepler-102 b	Transit		0.01353	0.042
Kepler-37 b	Transit		0.01	0.03

Appendix E

Working with the Densities of Exoplanets

For this activity will use the spreadsheet data available in Appendix D or at this book's IOP homepage, http://iopscience.iop.org/mono/978-0-7503-6290-0.

Part 1 Comparing exoplanet densities with densities of planets of similar size-classification in our own solar system.

    1 Convert the units of the MASS and RADII columns of the data from "Jupiters" to "Earths" (the necessary conversion factors are found in Table E1.

    2 Generate a DENSITY column (in units of Earth density) by dividing the MASS column by the cube of the RADIUS column.

Planet	Mass (Earth=1)	Radius (Earth=1)	Density (Earth=1)
Mercury	0.06	0.38	0.98
Venus	0.82	0.95	0.95
Earth	1.00	1.00	1.00
Mars	0.11	0.53	0.71
Jupiter	318.26	11.21	0.24
Saturn	95.14	9.45	0.13
Uranus	14.54	4.01	0.23
Neptune	17.09	3.88	0.30

Table E1. Solar system data.

doi:10.1088/2514-3433/ad7c3ech11   E-1   © IOP Publishing Ltd 2024. All rights,
including for text and data mining (TDM), artificial intelligence (AI) training, and similar technologies, are reserved.

3 Select out the necessary data to create a four (4) column table with the headings;

pl_name	MASS (E)	RAD (E)	DEN (E)

4 Now sort your data according to increasing radius.
5 Use Table E2 as guide to separate your data table into separate data tables for each type of exoplanet.
6 Sort each of your separate data tables by increasing mass and remove any *Earths* or *Super-Earths* greater than 10 Earth masses, any *Neptunes* greater than 50 Earth masses and any Jupiters or *larger* of more than 5000 Earth masses. These are the upper mass limits of each exoplanet category identified at https://www.space.com/36935-planet-classification.html.
7 Record the number of each planet type left and the percentage of the total number of exoplanets in Table E2.
8 Plot a histogram or bar graphing comparing the number of each exoplanet type.
9 Recombine all five or your data tables with their remaining exoplanets into one large data table. Generate a column for the LOG(DENSITY), the logarithm of the density and make a plot of the LOG(DENSITY) as a function of the RADIUS of all the exoplanets in your data set. Plotting the logarithm of the density instead of the density makes the plot simpler to analyze.
10 Does the general trend of your plot, the different densities of the different exoplanet types, generally agree with the densities of the planets in our own solar system from Table E1? Explain your answer.

Planet Type	Radius (Earths)	Number	% of Total
LARGER	>15		
JUPITER (Gas Giant)	6 to 15		
NEPTUNE (Ice Giant)	2 to 6		
SUPER EARTH	1.25 to 2		
EARTH LIKE	<1.25		

Table E2. Exoplanet types classified by size.

11 Select out *Earth* & *Super-Earth* data together from your data table and sort it according to increasing density. Most *Earths* and *Super-Earths* are expected to be composed mostly of rock and metal, so should have densities at least that of our solar system's Mars (see Table E1). Examine your data and determine how many of the total number of *Earths* and *Super-Earths* in your data table are of expected densities for *Earths* and *Super-Earths.* Also calculate the percentage. Can you think of anything that could explain their density? Hint—ALL the exoplanets in this data set are VERY close to their stars.

12 Repeat the previous question for your *Jupiters and larger* data, keeping in mind that these exoplanets are expected to be gas and liquid giants with densities similar to our solar system's Jupiter.

Part 2 Distinguishing between *Super-Earths, Mini-Neptunes and Neptunes.*

1 Select out and sort your *Super-Earth* data on increasing density, Any *Super-Earth* that is less dense than Mars (see Table E1) could actually be a *Mini-Neptune*. Select out all *Super-Earths* less dense that Mars and use it to start a new data table.

2 Select out and sort your *Neptune* data by increasing mass. The upper limit for Mini-*Neptunes* is considered 20 Earth masses. Select out all *Neptunes* less than this mass and combine them with the *Super-Earths* in your new data table.

3 Plot the density as a function of radius from your new data table.

4 At about what density does there appear to be a gap between *Super-Earths* and *Mini-Neptunes?*

5 At about what radius does there seem to be large drop in density? This should be the boundary between *Mini-Neptunes* and *Neputnes.*

6 Based on the last two questions identify the approximate density and radius boundaries for *Mini-Neptunes*.

7 Since *Super-Earths* are expected to be composed mostly of rock and metal and *Neptunes* more of gas and ices, what does the density range for *Mini-Neptunes* suggest about their composition?

8 Are there any exoplanets in your plot that don't appear to be a *Super-Earth, Mini-Neptune or Neptune?* Describe them in terms of their size and density compared to the rest of the exoplanets in your plot. Can you think of anything that could explain their size and density? Hint—ALL the exoplanets in this data set are VERY close to their stars.

Part 3 General Question.

Do all the exoplanets for which you have examined data fit into the size classifications (including *Mini-Neptunes*) listed in Table E1? If not, give examples and explain the reasons for them.

Student Exoplanet Projects Using Data from the Kepler Mission

Michael C LoPresto

Appendix F

TESS Spreadsheet Data

This appendix is also available as a spreadsheet from [doi link 10.1088/978-0-7503-6290-0]

doi:10.1088/2514-3433/ad7c3ech12

pl_name	hostname	pl_orbper	pl_orbsmax	pl_rade	pl_radj	pl_masse	pl_massj	st_spectype	st_teff	st_rad	st_mass	sy_dist
AU Mic b	AU Mic	8.4629991	0.0645			20.12	0.0633				0.5	9.7221
AU Mic c	AU Mic	18.858991	0.1101			9.6	0.0302				0.5	9.7221
BD-14 3065 b	BD-14 3065 A	4.2889731	0.0656	21.59	1.926	3932	12.37		6935	2.35	1.41	589.423
DS Tuc A b	DS Tuc A	8.138268		5.7	0.509			G6 V	5428	0.96	1.01	44.0622
GJ 1252 b	GJ 1252	0.5182331		1.193	0.106	1.32	0.00415	M3	3458	0.39	0.38	20.373
GJ 143 b	GJ 143	35.61253	0.1915	2.61	0.233	22.7	0.07142	K4.5	4640	0.69	0.73	16.32
GJ 3090 b	GJ 3090	2.8531054	0.03165	2.13	0.19	3.34	0.01051	M2 V	3556	0.52	0.52	22.4751
GJ 3473 b	GJ 3473	1.1980035	0.01589	1.264	0.113	1.86	0.00585		3347	0.36	0.36	27.3644
GJ 357 b	GJ 357	3.9306	0.036	1.2	0.107			M2.5 V	3505	0.34	0.34	9.44181
GJ 367 b	GJ 367	0.3219225	0.00709	0.699	0.062	0.633	0.00199	M1.0 V	3522	0.46	0.46	9.41263
GJ 3929 b	GJ 3929	2.616235	0.0252	1.09	0.097	1.75	0.00551		3384	0.32	0.31	15.8095
GJ 806 b	GJ 806	0.9263237	0.01406	1.331	0.119	1.9	0.00598	M1.5 V	3600	0.41	0.41	12.0445
Gliese 12 b	Gliese 12	12.761408	0.0668	0.958	0.085	3.87	0.01218	M3.0 V	3296	0.26	0.24	12.21
HD 108236 b	HD 108236	3.795963	0.04527	1.615	0.144				5660	0.88	0.87	64.5978
HD 108236 c	HD 108236	6.203449	0.062	2.071	0.185				5660	0.88	0.87	64.5978
HD 108236 d	HD 108236	14.175685	0.1074	2.539	0.227				5660	0.88	0.87	64.5978
HD 108236 e	HD 108236	19.590025	0.1367	3.083	0.275				5660	0.88	0.87	64.5978
HD 109833 b	HD 109833	9.188526		2.888	0.258				5881	1	1.08	79.5561
HD 109833 c	HD 109833	13.900148		2.59	0.231				5881	1	1.08	79.5561
HD 110082 b	HD 110082	10.18271	0.113	3.2	0.285	4.55	0.01432	F8 V	6200	1.19	1.21	105.096
HD 110113 b	HD 110113	2.541	0.035	2.05	0.183	131.89945	0.415	G	5732	0.97	1	106.307
HD 1397 b	HD 1397	11.53533	0.1097	11.5	1.026				5521	2.34	1.32	79.5702
HD 152843 b	HD 152843	11.6264	0.1053	3.41	0.304	11.56	0.03637	G0	6310	1.43	1.15	107.898
HD 152843 c	HD 152843	24.38		5.83	0.52	27.5	0.08652	G0	6310	1.43	1.15	107.898
HD 15337 b	HD 15337	4.75615	0.0522	1.64	0.146	7.51	0.02363	K1 V	5125	0.86	0.9	44.8155
HD 15337 c	HD 15337	17.1784	0.1268	2.39	0.213	8.11	0.02552	K1 V	5125	0.86	0.9	44.8155
HD 15906 b	HD 15906	10.924709	0.09	2.24	0.2			K V	4757	0.76	0.79	45.5621
HD 15906 c	HD 15906	21.583298	0.141	2.93	0.261			K V	4757	0.76	0.79	45.5621
HD 183579 b	HD 183579	17.471278	0.1334	3.55	0.317	19.7	0.06198		5788	0.98	1.03	57.2651
HD 18599 b	HD 18599	4.1374354	0.048	2.6	0.232	24.1	0.07583	K2 V	5145	0.77	0.86	38.5653
HD 191939 b	HD 191939	8.8803256	0.0804	3.41	0.304	10	0.03146	G9 V	5348	0.94	0.81	53.6089

HD 191939 c	HD 191939	28.579743	0.1752	3.195	0.285	8	0.02517	G9 V	5348	0.94	0.81	53.6089
HD 191939 d	HD 191939	38.353037	0.2132	2.995	0.267	2.8	0.00881	G9 V	5348	0.94	0.81	53.6089
HD 202772 A b	HD 202772 A	3.308958	0.05208	17.318	1.545	323.23311	1.017		6272	2.59	1.72	161.498
HD 20329 b	HD 20329	0.926118	0.018	1.72	0.153	7.42	0.02335		5596	1.13	0.9	63.6796
HD 207496 b	HD 207496	6.441008	0.0629	2.25	0.201	6.1	0.01919	K2.5 V	4819	0.77	0.8	23.606
HD 207897 b	HD 207897	16.20166	0.117	2.343	0.209	14.8	0.04657		5106	0.78	0.82	28.2938
HD 213885 b	HD 213885	1.008035	0.02012	1.745	0.156	8.83	0.02778	G	5978	1.1	1.07	48.0883
HD 21749 c	GJ 143	7.78993	0.0695	0.892	0.08	3.7	0.01164	K4.5	4640	0.69	0.73	16.32
HD 219666 b	HD 219666	6.04345	0.104	4.24	0.378	60.5	0.19	G5 V	5527	1.03	0.92	94.1719
HD 221416 b	HD 221416	14.2767	0.1228	9.17	0.836	26.57	0.0836		5080	2.94	1.21	95.5483
HD 22946 d	HD 22946	47.42489	0.2958	2.607	0.233				6169	1.12	1.1	62.7792
HD 23472 b	HD 23472	17.667087	0.1162	2	0.178	8.32	0.02618	K4 V	4684	0.71	0.67	39.0341
HD 23472 c	HD 23472	29.79749	0.1646	1.87	0.167	3.41	0.01073	K4 V	4684	0.71	0.67	39.0341
HD 23472 d	HD 23472	3.97664	0.04298	0.75	0.067	0.55	0.00173	K4 V	4684	0.71	0.67	39.0341
HD 23472 e	HD 23472	7.90754	0.068	0.818	0.073	0.72	0.00227	K4 V	4684	0.71	0.67	39.0341
HD 23472 f	HD 23472	12.1621839	0.0906	1.137	0.101	0.77	0.00242	K4 V	4684	0.71	0.67	39.0341
HD 235088 b	HD 235088	7.434151	0.071	2.075	0.185	4.2	0.01321	K4 V	5087	0.8	0.85	41.1706
HD 260655 b	HD 260655	2.76953	0.02933	1.24	0.111	2.14	0.00673	M0 V	3803	0.44	0.44	10.0055
HD 260655 c	HD 260655	5.70588	0.04749	1.533	0.137	3.09	0.00972	M0 V	3803	0.44	0.44	10.0055
HD 2685 b	HD 2685	4.12688	0.0568	16.141	1.44	371.8611	1.17		6801	1.56	1.43	196.852
HD 28109 b	HD 28109	22.89104	0.1357	2.199	0.196	18.496	0.05819	F8/G0 V	6120	1.45	1.26	139.611
HD 28109 c	HD 28109	56.00819	0.308	4.23	0.377	7.943	0.02499	F8/G0 V	6120	1.45	1.26	139.611
HD 28109 d	HD 28109	84.25999	0.411	3.25	0.29	5.681	0.01787	F8/G0 V	6120	1.45	1.26	139.611
HD 332231 b	HD 332231	18.712024	0.145	9.073	0.809	74.1	0.23314		6128	1.27	1.16	80.6665
HD 42813 b	HD 42813	13.630828	0.11	3.426	0.306	5.9	0.01856		5323	0.98	0.95	68.192
HD 5278 b	HD 5278	14.339156	0.1202	2.45	0.219	7.8	0.02454		6203	1.19	1.13	57.5846
HD 56414 b	HD 56414	29.04992	0.229	3.71	0.331			F V	8500	1.75	1.89	272.217
HD 63433 b	HD 63433	7.1079384	0.072	2.164	0.193	37.3	0.11736		5634	0.92	0.99	22.4035
HD 63433 c	HD 63433	20.543847	0.146	2.582	0.23	17.2	0.05412		5634	0.92	0.99	22.4035
HD 63433 d	HD 63433	4.209075	0.0503	1.073	0.096			G5 V	5688	0.91	0.99	22.4035
HD 63935 b	HD 63935	9.058807	0.083	2.916	0.26	10.5	0.03304		5513	0.96	0.94	48.9739
HD 73583 b	HD 73583	6.398042	0.0604	2.79	0.249	10.2	0.03209	K4 V	4511	0.65	0.73	31.5666

(Continued)

(Continued)

pl_name	hostname	pl_orbper	pl_orbsmax	pl_rade	pl_radj	pl_masse	pl_massj	st_spectype	st_teff	st_rad	st_mass	sy_dist
HD 73583 c	HD 73583	18.87974	0.1242	2.39	0.213	9.7	0.03052	K4 V	4511	0.65	0.73	31.5666
HD 86226 c	HD 86226	3.98442	0.049	2.16	0.193	7.25	0.02281	G1 V	5863	1.05	1.02	45.683
HD 93963 A b	HD 93963 A	1.0376109	0.02061	1.47	0.131	2.8	0.00881		5908	1.03	1.08	82.3432
HD 93963 A c	HD 93963 A	3.6451389	0.04726	3.12	0.278	18.4	0.05789		5908	1.03	1.08	82.3432
HIP 113103 b	HIP 113103	7.610303	0.06899	1.829	0.163				4930	0.74	0.76	46.2122
HIP 113103 c	HIP 113103	14.245648	0.10479	2.4	0.214				4930	0.74	0.76	46.2122
HIP 65 A b	HIP 65 A	0.9809734	0.01782	22.754	2.03	1021.18779	3.213	K4 V	4590	0.72	0.78	61.7856
HIP 67522 b	HIP 67522	6.959503		10.07	0.898				5675	1.38	1.22	127.28
HIP 94235 b	HIP 94235	7.713057	0.0787	3	0.268	379	1.19247	G V	5991	1.08	1.09	58.5146
HIP 9618 b	HIP 9618	20.772858	0.145	3.828	0.342	8.3	0.02611		5649	0.96	0.94	67.5478
HIP 9618 c	HIP 9618	52.56363	0.269	3.429	0.306	7	0.02202		5649	0.96	0.94	67.5478
HIP 97166 b	HIP 97166	10.288914	0.091	2.48	0.221	19.1	0.0601		5216	0.84	0.93	65.9501
HR 858 b	HR 858	3.58599	0.048	2.085	0.186				6201	1.31	1.15	31.964
HR 858 c	HR 858	5.97293	0.0674	1.939	0.173				6201	1.31	1.15	31.964
HR 858 d	HR 858	11.23	0.1027	2.164	0.193				6201	1.31	1.15	31.964
L 168-9 b	L 168-9	1.4015	0.02091	1.39	0.124	4.6	0.01447	M1 V	3800	0.6	0.62	25.1496
L 98-59 b	L 98-59	2.2531136	0.02191	0.85	0.076	0.4	0.00126	M3 V	3415	0.3	0.27	10.6194
L 98-59 c	L 98-59	3.6906777	0.0304	1.385	0.124	2.22	0.00698	M3 V	3415	0.3	0.27	10.6194
L 98-59 d	L 98-59	7.4507245	0.0486	1.521	0.136	1.94	0.0061	M3 V	3415	0.3	0.27	10.6194
LHS 1478 b	LHS 1478	1.9495378	0.01848	1.242	0.111	2.33	0.00733	m3 V	3381	0.25	0.24	18.2276
LHS 1678 b	LHS 1678	0.8602325	0.01239	0.685	0.061			M2.0 V	3490	0.33	0.34	19.8782
LHS 1678 c	LHS 1678	3.694284	0.0327	0.941	0.084			M2.0 V	3490	0.33	0.34	19.8782
LHS 1678 d	LHS 1678	4.9652229	0.04	0.981	0.088			M2.0 V	3490	0.33	0.34	19.8782
LHS 1815 b	LHS 1815	3.814334		1.088	0.097	1.58	0.00497	M1	3643	0.5	0.5	29.8424
LHS 3844 b	LHS 3844	0.46292913	0.00622	1.303	0.116			M5	3036	0.19	0.15	14.8846
LHS 475 b	LHS 475	2.029088		0.991	0.088			M3.5 V	3300	0.28	0.26	12.4814
LP 714-47 b	LP 714-47	4.052037	0.0417	4.7	0.419	30.8	0.09691	M0.0 V	3950	0.58	0.59	52.616
LP 791-18 b	LP 791-18	0.9479981	0.00978	1.212	0.108			M(6.1 +/- 0.7) V	2960	0.18	0.14	26.4927
LP 791-18 c	LP 791-18	4.9899093	0.02961	2.438	0.218	7.1	0.02234	M(6.1 +/- 0.7) V	2960	0.18	0.14	26.4927

Planet	Star											
LP 890-9 b	LP 890-9	2.7299025	0.01875	1.32	0.118	13.2	0.04153	M6 V	2850	0.16	0.12	32.4298
LTT 1445 A b	LTT 1445 A	5.35876	0.022	1.18	0.105			M3.0	3340	0.27	0.26	6.86929
LTT 1445 A c	LTT 1445 A	3.1239035	0.02661	1.147	0.102	1.54	0.00485		3340	0.27	0.26	6.86929
LTT 3780 b	LTT 3780.00	0.76837931	0.01195	1.325	0.118	2.46	0.00774		3358	0.38	0.38	21.9814
LTT 3780 c	LTT 3780.00	12.252284	0.0757	2.39	0.213	8.04	0.0253		3358	0.38	0.38	21.9814
LTT 9779 b	LTT 9779.00	0.792052	0.01679	4.72	0.421	29.32	0.09225	G7 V	5443	0.95	0.77	80.4373
NGTS-11 b	NGTS-11	35.45533	0.201	9.158	0.817	109.33352	0.344	K	5050	0.83	0.86	190.422
NGTS-20 b	NGTS-20	54.18915	0.313	11.994	1.07	947.12865	2.98	G1 IV	5980	1.78	1.47	382.898
NGTS-27 b	NGTS-27	3.3704181	0.0446	15.648	1.396	188.47225	0.593		5700	1.77	1.07	942.122
NGTS-30 b	NGTS-30	98.29838	0.408	10.402	0.928	305.11527	0.96		5455	0.91	0.94	238.377
TIC 139270665 b	TIC 139270665	23.624	0.163	7.23	0.645	147.15455	0.463	G2	5844	1.02	1.03	189.875
TIC 172900988 b	TIC 172900988 Aa	200.452	0.90281	11.25	1.004	942	2.96386	F9	6050	1.38	1.24	250.956
TIC 237913194 b	TIC 237913194	15.168865	0.1207	12.52	1.117	617.22586	1.942	G3	5788	1.09	1.03	306.063
TIC 241249530 b	TIC 241249530	165.7719	0.641	13.294	1.186	1582.78547	4.98		6166	1.4	1.27	324.448
TIC 257060897 b	TIC 257060897	3.660028	0.051	16.701	1.49	212.94503	0.67			1.82	1.32	505.703
TIC 279401253 b	TIC 279401253	76.8	0.369	11.209	1	1951.46642	6.14		5951	1.06	1.13	285.584
TIC 365102760 b	TIC 365102760	4.21285367	0.0622	6.21	0.554	19.2	0.06041		4900	3.13	1.17	555.546
TIC 393818343 b	TIC 393818343	16.24921	0.1291	12.184	1.087	1379.37529	4.34		5756	1.09	1.08	93.7345
TIC 46432937 b	TIC 46432937	1.44044527	0.02065	13.316	1.188	1017.0509	3.2		3572	0.53	0.56	90.5369
TOI-1052 b	TOI-1052	9.139703	0.09103	2.87	0.256	16.9	0.05317		6146	1.26	1.2	129.804
TOI-1062 b	TOI-1062	4.11296	0.052	2.265	0.202	10.15	0.03194	G9 V	5328	0.84	0.94	82.1733
TOI-1062 c	TOI-1062	7.972	0.08					G9 V	5328	0.84	0.94	82.1733
TOI-1064 b	TOI-1064	6.443868	0.06152	2.587	0.231	13.5	0.04248		4734	0.73	0.75	68.0726

(Continued)

(Continued)

pl_name	hostname	pl_orbper	pl_orbsmax	pl_rade	pl_radj	pl_masse	pl_massj	st_spectype	st_teff	st_rad	st_mass	sy_dist
TOI-1064 c	TOI-1064	12.226574	0.09429	2.651	0.237	2.5	0.00787		4734	0.73	0.75	68.0726
TOI-1075 b	TOI-1075	0.6047328	0.01159	1.791	0.16	9.95	0.03131	K9 V/M0 V	3875	0.58	0.6	61.4592
TOI-1107 b	TOI-1107	4.0782387	0.0561	14.572	1.3	1064.72516	3.35	F6 V	6311	1.81	1.35	281.314
TOI-1130 b	TOI-1130	4.07445	0.04457	3.56	0.318	19.28	0.06066	K6-K7	4350	0.69	0.71	58.2609
TOI-1130 c	TOI-1130	8.350231	0.07191	13.32	1.188	325.69	1.02474	K6-K7	4350	0.69	0.71	58.2609
TOI-1135 b	TOI-1135	8.0277283	0.082337	9.341	0.832				5962.7	1.15	1.16	114.048
TOI-1136 b	TOI-1136	4.1727	0.05106			3.5	0.01101		5770	0.97	1.02	84.5362
TOI-1136 c	TOI-1136	6.2574	0.0669			6.32	0.01988		5770	0.97	1.02	84.5362
TOI-1136 d	TOI-1136	12.5199	0.1062			8.35	0.02627		5770	0.97	1.02	84.5362
TOI-1136 e	TOI-1136	18.801				6.07	0.0191		5770	0.97	1.02	84.5362
TOI-1136 f	TOI-1136	26.321	0.174			9.7	0.03052		5770	0.97	1.02	84.5362
TOI-1136 g	TOI-1136	39.545	0.229			5.6	0.01762		5770	0.97	1.02	84.5362
TOI-1173 b	TOI-1173	7.06456	0.071	9.019	0.805	28.3	0.08904		5415	0.94	0.94	132.353
TOI-1174 b	TOI-1174	8.953458	0.08	2.506	0.224	32	0.10068		5158	0.78	0.84	94.6375
TOI-1180 b	TOI-1180	9.686753	0.082	3.036	0.271	10	0.03146		4804	0.74	0.78	72.0238
TOI-1181 b	TOI-1181	2.10319365	0.0364	16.197	1.445	377	1.18617		6046	1.93	1.46	302.819
TOI-1184 b	TOI-1184	5.748432	0.056	2.41	0.215	6.8	0.0214		4617	0.7	0.73	58.5946
TOI-1194 b	TOI-1194	2.3106446	0.034	8.717	0.778	120	0.37756		5394	0.96	0.98	149.667
TOI-1199 b	TOI-1199	3.671463	0.04988	10.514	0.938	75.96099	0.239		5710	1.45	1.23	248.427
TOI-1201 b	TOI-1201	2.4919863	0.0287	2.415	0.215	6.28	0.01976	M2.0 V	3476	0.51	0.51	37.8871
TOI-122 b	TOI-122	5.07803	0.0392	2.72	0.243			M3 V	3403	0.33	0.31	62.118
TOI-1221 b	TOI-1221	91.68278	0.387	2.91	0.26	1112.39942	3.5		5592	1.03	0.93	138.411
TOI-1224 b	TOI-1224	4.1782745	0.0355	2.104	0.188				3326	0.4	0.4	37.3312
TOI-1224 c	TOI-1224	17.945466		2.884	0.257				3326	0.4	0.4	37.3312
TOI-1227 b	TOI-1227	27.36397	0.0886	9.572	0.854			M4.5V-M5V	3072	0.56	0.17	100.641
TOI-1231 b	TOI-1231	24.245586	0.1288	3.65	0.326	15.4	0.04845	M3 V	3553	0.48	0.48	27.6227
TOI-1235 b	TOI-1235	3.444714		1.69	0.151	6.69	0.02105	M0.5 V	3997	0.62	0.63	39.6345
TOI-1244 b	TOI-1244	6.4003	0.061	2.376	0.212	6.6	0.02077		4722	0.72	0.75	102.551
TOI-1246 b	TOI-1246	4.30744	0.049	3.01	0.269	8.1	0.02549	K V	5151	0.86	0.87	169.422
TOI-1246 c	TOI-1246	5.904137	0.061	2.451	0.219	9.1	0.02863		5213	0.86	0.89	169.422
TOI-1246 d	TOI-1246	18.654874	0.132	3.431	0.306	5.4	0.01699		5213	0.86	0.89	169.422

TOI-1246 e	TOI-1246	37.92548	0.212	3.514	0.314	14.5	0.04562		5213	0.86	0.89	169.422
TOI-1247 b	HD 135694	15.92346	0.12	2.532	0.226	6.1	0.01919		5698	1.07	0.91	73.8713
TOI-1248 b	TOI-1248	4.3601561	0.051	6.808	0.607	27.4	0.08621		5205	0.87	0.9	168.735
TOI-1249 b	TOI-1249	13.079151	0.109	3.271	0.292	11.9	0.03744		5497	0.98	1.01	139.49
TOI-125 b	TOI-125	4.65382	0.05186	2.726	0.243	9.5	0.02989	KO V	5320	0.85	0.86	111.059
TOI-125 c	TOI-125	9.15059	0.0814	2.759	0.246	6.63	0.02086	KO V	5320	0.85	0.86	111.059
TOI-125 d	TOI-125	19.98	0.137	2.93	0.261	13.6	0.04279	KO V	5320	0.85	0.86	111.059
TOI-1259 A b	TOI-1259 A	3.477978	0.0407	11.456	1.022	140.16233	0.441		4775	0.71	0.74	118.106
TOI-1260 b	TOI-1260	3.127463	0.0367	2.41	0.215	8.56	0.02693		4227	0.67	0.68	73.5977
TOI-1260 c	TOI-1260	7.493134	0.0657	2.76	0.246	13.2	0.04153		4227	0.67	0.68	73.5977
TOI-1260 d	TOI-1260	16.608164	0.1116	3.12	0.278	11.84	0.03725		4227	0.67	0.68	73.5977
TOI-1266 b	TOI-1266	10.894841	0.0728	2.62	0.234	4.23	0.01331	M3 V	3618	0.44	0.43	36.0118
TOI-1266 c	TOI-1266	18.801611	0.1047	2.13	0.19	2.88	0.00906	M3 V	3618	0.44	0.43	36.0118
TOI-1268 b	TOI-1268	8.1577094	0.0711	9.1	0.812	96.4	0.30331	K1-K2	5300	0.92	0.96	109.557
TOI-1269 b	TOI-1269	4.2529913	0.0496	2.401	0.214	6.4	0.02014		5499	0.85	0.9	172.138
TOI-1272 b	TOI-1272	3.31599	0.0417	4.132	0.369	27	0.08495		5065	0.79	0.88	137.572
TOI-1273 b	TOI-1273	4.631296	0.0549	11.097	0.99	70.55791	0.222		5690	1.06	1.06	177.118
TOI-1278 b	TOI-1278	14.47567	0.095	12.218	1.09	5879.82553	18.5	M0 V	3799	0.57	0.54	75.5759
TOI-1279 b	TOI-1279	9.61419	0.085	2.661	0.237	10.6	0.03335		5457	0.84	0.88	107.055
TOI-128.01	TOI-128	4.9404681	0.05381	2.219	0.198				6086	1.13	0.85	68.4727
TOI-1288 b	TOI-1288	2.6998279	0.0374	4.968	0.443	44.1	0.13875		5388	0.96	0.96	114.865
TOI-1294 b	TOI-1294	3.915292	0.051	10.124	0.903	63.9	0.20105		5718	1.56	1.16	332.648
TOI-1294 c	TOI-1294	159.9	0.61						5718	1.56	1.16	332.648
TOI-1296 b	TOI-1296	3.9443736	0.051	14.116	1.259	95.3	0.29985		5567	1.7	1.16	323.265
TOI-1298 b	TOI-1298	4.537143	0.057	9.658	0.862	99	0.31149		5752	1.45	1.22	319.036
TOI-132 b	TOI-132	2.1097019	0.026	3.42	0.305	22.4	0.07048	G8 V	5397	0.9	0.97	163.678
TOI-1333 b	TOI-1333	4.720219	0.0626	15.648	1.396	753.2571	2.37		6274	1.93	1.46	200.492
TOI-1338 b	TOI-1338 A	95.174	0.4607	6.85	0.611	33	0.10383		6050	1.33	1.13	399.017
TOI-1347 b	TOI-1347	0.84742346		1.8	0.161	11.1	0.03492		5464	0.83	0.91	147.476
TOI-1347 c	TOI-1347	4.841962		1.6	0.143	6.4	0.02014		5464	0.83	0.91	147.476
TOI-1386 b	TOI-1386	25.8401	0.173	6.216	0.555	45.8	0.1441		5735	1.02	1.04	146.858
TOI-139 b	TOI-139	11.07085		2.457	0.219				4570	0.7	0.69	42.4061

(Continued)

(Continued)

pl_name	hostname	pl_orbper	pl_orbsmax	pl_rade	pl_radj	pl_masse	pl_massj	st_spectype	st_teff	st_rad	st_mass	sy_dist
TOI-1408 b	TOI-1408	4.424711	0.05804	16.813	1.5	537.13001	1.69		6306	1.27	1.33	139.484
TOI-1410 c	TOI-1410	47.56	0.239						4635	0.76	0.79	72.7512
TOI-1410.01	TOI-1410	1.21687	0.0207	3.101	0.277	12.5	0.03933		4635	0.76	0.79	72.7512
TOI-1411 b	TOI-1411	1.4520527	0.02122	1.199	0.107	2	0.00629		4115	0.61	0.6	32.4778
TOI-1416 b	TOI-1416	1.0697568	0.019	1.62	0.145	3.48	0.01095	G9 V	4884	0.79	0.8	55.0135
TOI-1420 b	TOI-1420	6.9561063	0.071	11.89	1.061	25.1	0.07897		5510	0.92	0.99	201.918
TOI-1422 b	TOI-1422	12.9972	0.108	3.96	0.353	9	0.02832	G2 V	5840	1.02	0.98	155.141
TOI-1431 b	TOI-1431	2.650237	0.046	16.701	1.49	991.62463	3.12	Am C	7690	1.92	1.9	148.925
TOI-1437 b	TOI-1437	18.84078	0.139	2.435	0.217	10.4	0.03272		5986	1.26	1.01	103.297
TOI-1439 b	TOI-1439	27.644	0.192	4.243	0.379	38.5	0.12113		5841	1.62	1.23	230.306
TOI-1442 b	TOI-1442	0.4090677	0.0071	1.17	0.104				3330	0.31	0.29	41.1659
TOI-1443 b	TOI-1443	23.540678	0.147	2.304	0.206	30	0.09439		5224	0.75	0.77	85.8034
TOI-1444 b	TOI-1444	0.4702743	0.0116	1.418	0.126	3.58	0.01126		5460	0.91	0.95	125.463
TOI-1448 b	TOI-1448	8.112245	0.0567	2.749	0.245	19.5	0.06135		3412	0.38	0.37	73.3345
TOI-1450 A b	TOI-1450 A	2.0439274		1.13	0.101	1.258	0.00396	M3.0 V	3437	0.48	0.48	22.4467
TOI-1451 b	TOI-1451	16.537944	0.127	2.611	0.233	15.2	0.04782		5801	1.02	1	91.9362
TOI-1452 b	TOI-1452	11.06201	0.061	1.672	0.149	4.82	0.01517	M4 &#177 0.5 V	3185	0.28	0.25	30.5212
TOI-1467 b	TOI-1467	5.97199	0.0495	1.676	0.15	9	0.02832		3776	0.46	0.45	37.4418
TOI-1468 b	TOI-1468	1.8805136		1.28	0.114	3.21	0.0101	M3.0 V	3496	0.34	0.34	24.7399
TOI-1468 c	TOI-1468	15.532482		2.064	0.184	6.64	0.02089	M3.0 V	3496	0.34	0.34	24.7399
TOI-1470 b	TOI-1470	2.527093	0.0285	2.18	0.194	7.32	0.02303	M1.5 V	3709	0.47	0.47	51.9503
TOI-1470 c	TOI-1470	18.08816	0.106	2.47	0.22	7.24	0.02278	M1.5 V	3709	0.47	0.47	51.9503
TOI-1472 b	TOI-1472	6.36381	0.065	4.16	0.371	16.5	0.05191		5133	0.84	0.92	121.84
TOI-1473 b	HD 6061	5.2549	0.06	2.428	0.217	10	0.03146		5934	1.01	1.03	67.6203
TOI-1478 b	TOI-1478	10.180249	0.0903	11.882	1.06	270.47333	0.851		5597	1.05	0.95	152.892
TOI-150.01	TOI-150	5.857487	0.07037	14.067	1.255	797.7533	2.51	F	6255	1.53	1.35	336.267
TOI-1516 b	TOI-1516	2.056014		15.244	1.36	1004.33777	3.16	F8 V	6520	1.14	1.14	247.054
TOI-157 b	TOI-157	2.0845435	0.03138	14.415	1.286	375.0394	1.18	G9 IV	5404	1.17	0.95	355.683
TOI-1601 b	TOI-1601	5.33175	0.069	14.199	1.267	387	1.21764		5982	2.19	1.51	336.608
TOI-163 b	TOI-163	4.231306	0.058	16.69	1.489	387.7526	1.22	F	6495	1.65	1.44	411.719

Planet	Star											
TOI-1634 b	TOI-1634	0.9893455		1.773	0.158	7.57	0.02382		3472	0.45	0.45	35.2736
TOI-1669 c	TOI-1669	2.6800535	0.0376	2.296	0.205	13	0.0409		5497	1.06	0.99	111.276
TOI-1670 b	TOI-1670	10.98462	0.103	2.06	0.184	41.31769	0.13	F7 V	6170	1.32	1.21	168.067
TOI-1670 c	TOI-1670	40.74976	0.249	11.063	0.987	200.2319	0.63	F7 V	6170	1.32	1.21	168.067
TOI-1680 b	TOI-1680	4.8026345	0.03144	1.466	0.131			M4.5+/-0.5	3225	0.21	0.18	37.2209
TOI-1683.01	TOI-1683	3.0575356	0.036508	2.638	0.235				4402	0.7	0.69	51.1884
TOI-1685 b	TOI-1685	0.6691406		1.7	0.152	3.09	0.00972	M3.0 V	3434	0.49	0.49	37.6153
TOI-169 b	TOI-169	2.2554477	0.03524	12.173	1.086	251.40353	0.791	G1 V	5880	1.29	1.15	407.84
TOI-1691 b	TOI-1691	16.7369	0.126	3.565	0.318	14.6	0.04594		5642	1.01	0.96	111.445
TOI-1693 b	TOI-1693	1.7666957	0.0226	1.41	0.126				3499	0.46	0.49	30.7947
TOI-1694 b	TOI-1694	3.770107	0.045	5.342	0.477	31.3	0.09848		5058	0.8	0.85	124.68
TOI-1695 b	TOI-1695	3.1342791	0.033548	1.9	0.17	6.36	0.02001	M1 V	3690	0.52	0.51	45.1309
TOI-1696 b	TOI-1696	2.500311	0.0229	3.09	0.276	48.8	0.15354	M5V	3185	0.28	0.26	64.9172
TOI-1710 b	TOI-1710	24.283377	0.166	5.203	0.464	22.4	0.07048		5684	0.96	1.03	81.2271
TOI-172 b	TOI-172	9.47725	0.0914	10.817	0.965	1722.6386	5.42		5645	1.78	1.13	342.831
TOI-1723 b	TOI-1723	13.72641	0.114	3.292	0.294	10.4	0.03272		5743	1.09	1.04	100.727
TOI-1728 b	TOI-1728	3.49151	0.0391	5.05	0.451	26.78	0.08426	M0	3980	0.62	0.65	60.798
TOI-1736 b	TOI-1736	7.073076	0.073	3.182	0.284	12.3	0.0387		5636	1.43	1.04	88.9482
TOI-1739 b	TOI-1739	8.303342	0.0742	1.695	0.151				4922	0.75	0.79	70.9819
TOI-1742 b	TOI-1742	21.269084	0.154	2.365	0.211	9.7	0.03052		5815	1.13	1.09	72.8871
TOI-1749 b	TOI-1749	2.38839	0.0291	1.39	0.124	57	0.17934	M0 V	3985	0.55	0.58	99.5561
TOI-1749 c	TOI-1749	4.4929	0.0443	2.12	0.189	14	0.04405	M0 V	3985	0.55	0.58	99.5561
TOI-1749 d	TOI-1749	9.0497	0.0707	2.52	0.225	15	0.0472	M0 V	3985	0.55	0.58	99.5561
TOI-1751 b	TOI-1751	37.46852	0.215	2.951	0.263	19.5	0.06135		5970	1.31	0.94	113.335
TOI-1753 b	TOI-1753	5.3846104	0.059	2.479	0.221	16.6	0.05223		5621	0.97	0.95	226.671
TOI-1758 b	TOI-1758	20.705127	0.138	3.557	0.317	6.9	0.02171		5151	0.82	0.83	96.6215
TOI-1759 b	TOI-1759	18.85019	0.1177	3.14	0.28	10.8	0.03398	M0.0 V	4065	0.6	0.61	40.0654
TOI-1775 b	TOI-1775	10.2405549	0.08	8.047	0.718	96	0.30205		5284	0.83	0.92	149.234
TOI-1776 b	TOI-1776	2.799868	0.0378	1.216	0.108	1.4	0.0044		5785	0.94	0.92	44.647
TOI-1778 b	HD 77946	6.527363	0.073	2.896	0.258	10.9	0.0343		6007	1.32	1.2	99.451
TOI-178 b	TOI-178	1.914558	0.02607	1.152	0.103	1.5	0.00472	K	4316	0.65	0.65	62.699
TOI-178 c	TOI-178	3.23845	0.037	1.669	0.149	4.77	0.01501	K	4316	0.65	0.65	62.699

(Continued)

(Continued)

pl_name	hostname	pl_orbper	pl_orbsmax	pl_rade	pl_radj	pl_masse	pl_massj	st_spectype	st_teff	st_rad	st_mass	sy_dist
TOI-178 d	TOI-178	6.5577	0.0592	2.572	0.229	3.01	0.00947	K	4316	0.65	0.65	62.699
TOI-178 e	TOI-178	9.961881	0.0783	2.207	0.197	3.86	0.01214	K	4316	0.65	0.65	62.699
TOI-178 g	TOI-178	20.7095	0.1275	2.87	0.256	3.94	0.0124	K	4316	0.65	0.65	62.699
TOI-1789 b	TOI-1789	3.208664	0.04882	16.141	1.44	222.47988	0.7	F	5991	2.17	1.51	229.067
TOI-1794 b	TOI-1794	8.765528	0.082	3.297	0.294	8.7	0.02737		5631	1.32	0.97	155.791
TOI-1798 c	TOI-1798	0.4378146	0.0107	1.399	0.125	5.6	0.01762		5106	0.79	0.85	113.181
TOI-1798.01	TOI-1798	8.02154	0.074	2.385	0.213	6.5	0.02045		5106	0.79	0.85	113.181
TOI-1799 b	TOI-1799	7.085738	0.071	1.422	0.127	4	0.01259		5698	0.95	0.94	62.1274
TOI-1801 b	TOI-1801	10.64387		2.079	0.186	5.738	0.01805	M0.5+/-0.5 V	3863	0.55	0.56	30.6796
TOI-1806.01	TOI-1806	15.1454693	0.087735	3.41	0.304				3272	0.4	0.39	55.5211
TOI-1807 b	TOI-1807	0.54937097	0.0122	1.497	0.134	2.44	0.00768		4914	0.75	0.8	42.5775
TOI-181 b	TOI-181	4.532	0.0539	6.956	0.634	46.1687	0.1452	K	4994	0.74	0.82	96.2338
TOI-1811 b	TOI-1811	3.7130765	0.04389	11.142	0.994	308.92921	0.972		4766	0.77	0.82	128.23
TOI-1820 b	TOI-1820	4.860674	0.061	12.778	1.14	731.00534	2.3		5734	1.51	1.04	248.504
TOI-1823 b	TOI-1823	38.8	0.212	7.541	0.673	67.4	0.21206		4927	0.8	0.84	71.6456
TOI-1824 b	TOI-1824	22.808533	0.151	2.744	0.245	16	0.05034		5165	0.81	0.88	59.4314
TOI-1836 b	TOI-1836	20.380845	0.157	8.281	0.739	24.7	0.07771		6237	1.65	1.25	191.649
TOI-1836 c	TOI-1836	1.7727471	0.03089	2.6	0.232	8	0.02517		6237	1.65	1.25	191.649
TOI-1842 b	TOI-1842	9.5739181	0.1	12.349	1.102	89.4	0.28128		6039	2.03	1.46	223.465
TOI-1853 b	TOI-1853	1.2436258	0.0213	3.46	0.309	73.2	0.23031	K2.5 V	4985	0.81	0.84	165.105
TOI-1855 b	TOI-1855	1.36414864	0.02398	18.495	1.65	360.09959	1.133		5359	1.04	0.99	176.833
TOI-1859 b	TOI-1859	63.48347	0.337	9.752	0.87				6341	1.36	1.29	224.385
TOI-1860 b	TOI-1860	1.0662107	0.0204	1.31	0.117			G V	5752	0.94	0.99	45.864
TOI-1898 b	TOI-1898	45.52234	0.269	9.396	0.838	129	0.40588		6241	1.61	1.25	79.6682
TOI-1899 b	TOI-1899	29.090312	0.1525	11.097	0.99	212.94503	0.67		3926	0.61	0.63	128.438
TOI-1937 A b	TOI-1937 A	0.94667944	0.01932	13.978	1.247	638.8351	2.01		5814	1.08	1.07	415.135
TOI-198 b	TOI-198	10.2152	0.1	1.44	0.128			M0 V	3650	0.44	0.47	23.7262
TOI-199 b	TOI-199	104.854	0.4254	9.079	0.81	54.03083	0.17	G9 V	5255	0.82	0.94	102.27
TOI-1994 b	TOI-1994	4.0337142	0.0613	13.675	1.22	7024.00779	22.1		7700	2.3	1.86	515.603
TOI-2000 b	TOI-2000	3.09833	0.04271	2.7	0.241	11	0.0347		5611	1.13	1.08	175.713
TOI-2000 c	TOI-2000	9.127055	0.0878	8.14	0.727	81.7	0.257		5611	1.13	1.08	175.713

TOI-201 b	52.97818	0.3	11.299	1.008	133.4886	0.42	F6 V	6394	1.32	1.32	113.825
TOI-2010 b	141.834025	0.5516	11.814	1.054	408.72733	1.286	F0	5929	1.08	1.11	109.19
TOI-2015 b	3.348968	0.0301	3.39	0.302	16.4	0.0516	M4	3194	0.33	0.34	47.3444
TOI-2018 b	7.435583		2.268	0.202	9.2	0.02895	K V	4174	0.62	0.57	27.9956
TOI-2019 b	15.3444	0.128	5.333	0.476	34.4	0.10823		5626	1.75	1.18	198.007
TOI-2025 b	8.8720982	0.0892	11.893	1.061	1144.18227	3.6		5977	1.46	1.2	335.985
TOI-2046 b	1.4971842		16.141	1.44	731.00534	2.3	F8 V	6250	1.21	1.13	289.668
TOI-2048 b	13.79019		2.61	0.233			G V	5185	0.79	0.83	116.567
TOI-206 b	0.7363104	0.0112	1.3	0.116				3383	0.35	0.35	47.7465
TOI-2068 b	7.768915	0.0632	1.821	0.162				3710	0.54	0.56	52.9328
TOI-2076 b	10.355096	0.088	2.771	0.247	16.1	0.05066		5192	0.8	0.86	41.9091
TOI-2076 c	21.015327	0.142	3.694	0.33	55	0.17305		5192	0.8	0.86	41.9091
TOI-2076 d	35.125686	0.199	3.431	0.306	29	0.09124		5192	0.8	0.86	41.9091
TOI-2081 b	10.50534	0.0752	2.04	0.182			M1.0	3800	0.53	0.54	62.3446
TOI-2084 b	6.0784247	0.05006	2.47	0.22			M2+/-0.5	3551	0.47	0.45	114.549
TOI-2088 b	124.72997	0.472	3.683	0.329	37	0.11642		5081	0.85	0.9	126.46
TOI-2095 b	17.66484	0.101	1.25	0.112	4.1	0.0129	M2.5 V	3759	0.44	0.44	41.9176
TOI-2095 c	28.17232	0.137	1.33	0.119	7.4	0.02328	M2.5 V	3759	0.44	0.44	41.9176
TOI-2096 b	3.1190633	0.025	1.243	0.111			M4	3300	0.23	0.23	48.4809
TOI-2096 c	6.38784	0.04	1.914	0.171			M4	3300	0.23	0.23	48.4809
TOI-2107 b	2.4545467	0.03515	13.574	1.211	263.79758	0.83		5627	0.93	0.96	236.705
TOI-2109 b	0.67247414	0.01791	15.098	1.347	1595.4986	5.02	F	6540	1.7	1.45	262.041
TOI-2120 b	5.7998164	0.0377	2.122	0.189	6.8	0.0214		3131	0.24	0.21	32.1755
TOI-2128 b	16.34136	0.127	2.089	0.186	4.5	0.01416		5968	1.12	1.03	36.578
TOI-2134 b	9.2292005	0.078	2.69	0.24	9.13	0.02873	K5 V	4580	0.71	0.74	22.6202
TOI-2134 c	95.5	0.371	7.27	0.649	41.89	0.1318	K5 V	4580	0.71	0.74	22.6202
TOI-2136 b	7.851928	0.057	2.19	0.195	6.37	0.02004	M4.5 V	3342	0.34	0.34	33.3631
TOI-2141 b	18.26157	0.133	3.05	0.278	24	0.075		5659	0.98	0.94	77.682
TOI-2145 b	10.261127	0.111	12.419	1.108	1767	5.5596		6200	2.75	1.72	224.731
TOI-2152 A b	3.3773512	0.05064	14.359	1.281	899.45439	2.83	F4 V	6630	1.61	1.52	303.006
TOI-2154 b	3.8240801	0.0513	16.287	1.453	292.40213	0.92		6280	1.4	1.23	296.46
TOI-2158 b	8.60077	0.075	10.761	0.96	260.61929	0.82		5673	1.41	1.12	198.473

(Continued)

(Continued)

pl_name	hostname	pl_orbper	pl_orbsmax	pl_rade	pl_radj	pl_masse	pl_massj	st_spectype	st_teff	st_rad	st_mass	sy_dist
TOI-216.01	TOI-216	34.525528		10.1	0.901	177.98391	0.56			0.75	0.77	177.945
TOI-216.02	TOI-216	17.16073		8	0.714	18.75188	0.059			0.75	0.77	177.945
TOI-2180 b	TOI-2180	260.79	0.828	11.321	1.01	875.61726	2.755	G5	5695	1.64	1.11	116.685
TOI-2184 b	TOI-2184	6.90683		11.4	1.017	206.58846	0.65		5966	2.9	1.53	788.317
TOI-2193 A b	TOI-2193 A	2.1225735	0.03319	19.84	1.77	298.7587	0.94		5966	1.25	1.08	337.064
TOI-2194 b	TOI-2194	15.337597		1.989	0.177				4756	0.69	0.74	19.5711
TOI-2196 b	TOI-2196	1.1947268	0.02234	3.51	0.313	26	0.08181	G V	5634	1.04	1.03	259.947
TOI-220 b	TOI-220	10.695264	0.08911	3.03	0.27	13.8	0.04342	K0 V	5298	0.86	0.82	90.7402
TOI-2202 b	TOI-2202	11.9101	0.09564	11.321	1.01	310.83618	0.978	K8 V	5144	0.79	0.82	235.933
TOI-2207 b	TOI-2207	8.001968	0.0854	11.153	0.995	203.41018	0.64		6101	1.56	1.3	373.345
TOI-2236 b	TOI-2236	3.5315902	0.05009	14.37	1.282	502.16888	1.58		6248	1.57	1.34	355.59
TOI-2257 b	TOI-2257	35.189346	0.145	2.194	0.196			M3	3430	0.31	0.33	57.7911
TOI-2260 b	TOI-2260	0.3524728	0.0097	1.62	0.145			G V	5534	0.94	0.99	101.251
TOI-2266 b	TOI-2266	2.326318	0.02	1.54	0.137			M5.0+0.5-0.5	3200	0.24	0.23	51.6004
TOI-2285 b	TOI-2285	27.26955	0.1363	1.74	0.155	19.5	0.06135		3491	0.46	0.45	42.409
TOI-2337 b	TOI-2337	2.99432		10.088	0.9	508.52545	1.6		4780	3.22	1.32	536.046
TOI-2338 b	TOI-2338	22.65398	0.158	11.209	1	1900.61387	5.98		5581	1.05	0.99	313.996
TOI-2364 b	TOI-2364	4.0197517	0.04871	8.608	0.768	71.51139	0.225		5306	0.89	0.95	219.614
TOI-2368 b	TOI-2368	5.1750073	0.05649	10.839	0.967	206.58846	0.65		5360	0.85	0.9	210.915
TOI-237 b	TOI-237	5.436098	0.0341	1.44	0.128			M4.5 V	3212	0.21	0.18	38.0749
TOI-2373 b	TOI-2373	13.33668	0.112	10.424	0.93	2955.80419	9.3		5651	1.1	1.04	508.686
TOI-2374 b	TOI-2374	4.31361	0.0471	6.81	0.608	56.64	0.17821		4802	0.69	0.75	134.655
TOI-238 b	TOI-238	1.273114	0.0212	1.612	0.144				5080	0.75	0.79	80.5407
TOI-2406 b	TOI-2406	3.0766891	0.02267	2.86	0.255	15.6	0.04908		3100	0.2	0.17	55.419
TOI-2411 b	TOI-2411	0.7826942	0.0144	1.68	0.15				4099	0.68	0.65	59.5362
TOI-2416 b	TOI-2416	8.275479	0.0831	9.864	0.88	953.48522	3		5808	1.24	1.12	557.733
TOI-2421 b	TOI-2421	4.3474032	0.0543	10.368	0.925	105.83686	0.333		5607	1.75	1.13	327.927
TOI-2427 b	TOI-2427	1.3060011	0.0202	1.8	0.161				4072	0.65	0.64	28.5147
TOI-244 b	TOI-244	7.397225	0.0559	1.52	0.136	2.68	0.00843	M2.5 V	3433	0.43	0.43	22.0337
TOI-2443 b	TOI-2443	15.669494		2.773	0.247				4357	0.73	0.66	23.9258
TOI-2445 b	TOI-2445	0.3711281	0.0064	1.25	0.112				3333	0.27	0.25	48.582

TOI-2447 b	TOI-2447	69.33684	0.347	9.606	0.857	124.90656	0.393		5730	1.01	1.03	148.987
TOI-2459 b	TOI-2459	19.104718		2.953	0.263		4.82		4195	0.68	0.66	36.6113
TOI-2497 b	TOI-2497	10.655669	0.1166	11.142	0.994	1531.93292	0.10893		7360	2.36	1.86	285.289
TOI-2498 b	TOI-2498	3.738252	0.0491	6.06	0.541	34.62	1		5905	1.26	1.12	275.265
TOI-251 b	TOI-251	4.93777	0.05741	2.74	0.244	317.83	0.64		5875	0.88	1.04	99.523
TOI-2524 b	TOI-2524	7.18585	0.073	11.209	1	203.41018	0.084		5831	1.12	1.01	438.842
TOI-2525 b	TOI-2525	23.2856	0.1511	8.676	0.774	26.69759	0.657	K8 V	5096	0.79	0.85	395.398
TOI-2525 c	TOI-2525	49.2519	0.249	10.133	0.904	208.81326		K8 V	5096	0.79	0.85	395.398
TOI-2545 b	TOI-2545	7.994037		2.75	0.245				5846.29	1.25	1.05	107.118
TOI-2567 b	TOI-2567	5.983944	0.0672	10.929	0.975	63.88351	0.201		5611	1.69	1.13	504.775
TOI-257 b	TOI-257	18.38818	0.1528	7.163	0.639	43.86032	0.138		6095	1.87	1.41	76.9028
TOI-2570 b	TOI-2570	2.9887615	0.04145	13.641	1.217	260.61929	0.82		5771	1.09	1.06	361.857
TOI-2583 A b	TOI-2583 A	4.5207265	0.0571	14.46	1.29	79.4571	0.25		5936	1.48	1.22	566.189
TOI-2587 A b	TOI-2587 A	5.45664	0.0635	12.072	1.077	69.28659	0.218		5760	1.73	1.15	373.675
TOI-2589 b	TOI-2589	61.6277	0.3	12.106	1.08	1112.39942	3.5		5579	1.07	0.93	200.273
TOI-260 b	TOI-260	13.475815	0.091	1.473	0.131	3.3	0.01038		4050	0.55	0.55	20.1852
TOI-261.01	TOI-261	3.3639254	0.034792	3.045	0.271				5890.3	1.28	0.5	114.495
TOI-262 b	TOI-262	11.14529	0.163	2.07	0.185			K0 V	5310	0.85	0.91	43.932
TOI-2641 b	TOI-2641	4.880974	0.0607	18.103	1.615	122.68177	0.386	F9 V	6100	1.34	1.16	344.045
TOI-266 b	HIP 8152	10.751014	0.093	2.536	0.226	8.9	0.028		5618	0.96	0.94	101.694
TOI-266 c	HIP 8152	19.60562	0.139	2.524	0.225	10.7	0.03367		5618	0.96	0.94	101.694
TOI-2669 b	TOI-2669	6.2034		19.728	1.76	193.87533	0.61		4800	4.1	1.19	393.443
TOI-269 b	TOI-269	3.6977104	0.0345	2.77	0.247	8.8	0.02769	M2 V	3514	0.4	0.39	57.0225
TOI-270 b	TOI-270	3.3601538	0.03197	1.206	0.108	1.58	0.00497	M	3506	0.38	0.39	22.4793
TOI-270 c	TOI-270	5.6605731	0.04526	2.355	0.21	6.15	0.01935	M	3506	0.38	0.39	22.4793
TOI-270 d	TOI-270	11.379573	0.0721	2.133	0.19	4.78	0.01504	M	3506	0.38	0.39	22.4793
TOI-277 b	TOI-277	3.994	0.0269	2.65	0.236				4031	0.36	0.16	64.6692
TOI-2796 b	TOI-2796	4.8084983	0.0569	17.262	1.54	139.8445	0.44		5764	1.07	1.06	350.343
TOI-2803 A b	TOI-2803 A	1.96229325	0.03185	18.114	1.616	309.8827	0.975		6280	1.25	1.12	494.818
TOI-2818 b	TOI-2818	4.039709	0.0493	15.278	1.363	225.65817	0.71		5721	1.23	0.98	312.674
TOI-2842 b	TOI-2842	3.5514058	0.0475	12.845	1.146	117.59651	0.37		5910	1.26	1.14	460.995
TOI-2977 b	TOI-2977	2.3505614	0.03386	13.159	1.174	533.95172	1.68		5691	1.07	0.94	

(Continued)

(Continued)

pl_name	hostname	pl_orbper	pl_orbsmax	pl_rade	pl_radj	pl_masse	pl_massj	st_spectype	st_teff	st_rad	st_mass	sy_dist
TOI-3023 b	TOI-3023	3.9014971	0.0505	16.432	1.466	197.05361	0.62		5760	1.67	1.12	389.916
TOI-3071 b	TOI-3071	1.266938	0.0249	7.16	0.639	68.2	0.21458		6177	1.31	1.29	485.219
TOI-3082 b	TOI-3082	1.926907		3.662	0.327				4263	0.68	0.66	113.048
TOI-3235 b	TOI-3235	2.59261842	0.02709	11.4	1.017	211.35589	0.665	M V	3388.8	0.37	0.39	72.811
TOI-3261 b	TOI-3261	0.8831331	0.01714	3.82	0.341	30.3	0.09533	K1.5+/-1	5068	0.85	0.86	300.141
TOI-329 b	TOI-329	5.70439	0.0638	4.941	0.441	40.4	0.12711		5614	1.53	1.07	284.388
TOI-332 b	TOI-332	0.777038	0.0159	3.2	0.285	57.2	0.17997	K0 V	5251	0.87	0.88	222.852
TOI-3321 b	TOI-3321	3.6525145	0.047	15.558	1.388	176.07694	0.554		5850	1.55	1.04	285.895
TOI-3331 A b	TOI-3331 A	2.0180231	0.03144	12.98	1.158	721.47048	2.27		5609	0.95	1.02	217.193
TOI-3353.01	TOI-3353	4.6657965	0.060118	2.667	0.238				6365	1.02	1.33	78.7577
TOI-3362 b	TOI-3362	18.09547	0.153	12.801	1.142	1598.35906	5.029		6532	1.83	1.45	366.987
TOI-3364 b	TOI-3364	5.8768918	0.0675	12.229	1.091	530.77344	1.67		5706	1.42	1.19	277.899
TOI-3540 A b	TOI-3540 A	3.119999	0.04289	23.539	2.1	375.03752	1.18		5810	1.23	1.08	
TOI-3629 b	TOI-3629	3.936551	0.043	8.295	0.74	82.63539	0.26	M1	3870	0.6	0.63	130.091
TOI-3688 A b	TOI-3688 A	3.246075	0.0456	13.081	1.167	311.47184	0.98		5950	1.3	1.2	396.378
TOI-3693 b	TOI-3693	9.088516	0.0813	12.599	1.124	324.18498	1.02		5321	0.79	0.87	176.436
TOI-3714 b	TOI-3714	2.154849	0.027	11.321	1.01	222.47988	0.7	M2	3660	0.51	0.53	112.71
TOI-3757 b	TOI-3757	3.438753	0.03845	12	1.071	85.3	0.26838	M0 V	3913	0.62	0.64	181.139
TOI-3785 b	TOI-3785	4.6747373	0.043	5.14	0.459	14.95	0.04704	M2 V	3576	0.5	0.52	79.7634
TOI-3807 b	TOI-3807	2.8989727	0.0421	18.495	1.65	330.54154	1.04		5772	1.47	1.18	421.745
TOI-3819 b	TOI-3819	3.2443141	0.04611	13.137	1.172	352.78953	1.11		5859	1.54	1.24	558.141
TOI-3884 b	TOI-3884	4.5445697	0.0354	6	0.535	16.5	0.05191	M4	3269	0.3	0.28	43.1448
TOI-3894 b	TOI-3894	4.33454	0.05354	15.278	1.363	260.61929	0.82		6000	1.5	1.14	417.365
TOI-3912 b	TOI-3912	3.4936264	0.0463	14.28	1.274	129.03833	0.406		5725	1.39	1.09	465.992
TOI-3919 b	TOI-3919	7.433234	0.0795	12.319	1.099	1233.17422	3.88		6100	1.32	1.21	604.591
TOI-3976 A b	TOI-3976 A	6.607662	0.0743	12.274	1.095	55.61997	0.175		5975	1.5	1.25	516.087
TOI-3984 A b	TOI-3984 A	4.353326	0.041	7.9	0.71	44	0.14	M4+/-0.5	3476	0.47	0.49	108.883
TOI-4010 b	TOI-4010	1.348335	0.0229	3.02	0.269	11	0.03461	K V	4960	0.83	0.88	177.504
TOI-4010 c	TOI-4010	5.414654	0.058	5.93	0.529	20.31	0.0639	K V	4960	0.83	0.88	177.504
TOI-4010 d	TOI-4010	14.70886	0.113	6.18	0.551	38.15	0.12003	K V	4960	0.83	0.88	177.504
TOI-406.01	TOI-406	13.1757294	0.085518	1.96	0.175				3349	0.39	0.48	31.0597

TOI-4087 b	TOI-4087	3.1774835	0.04469	13.047	1.164	232.01474	0.73		6060	1.11	1.11	1.18	301.676
TOI-411 b	HD 22946	4.040295	0.05727	1.362	0.122	13.71	0.04314		6169	1.12	1.12	1.1	62.7792
TOI-411 c	HD 22946	9.573083	0.1017	2.328	0.208	9.72	0.03058		6169	1.12	1.12	1.1	62.7792
TOI-4127 b	TOI-4127	56.39879	0.3081	12.285	1.096	731.00534	2.3	F V	6096	1.29	1.29	1.23	319.632
TOI-4137 b	TOI-4137	3.8016122	0.05222	13.574	1.211	457.67291	1.44		6202	1.44	1.44	1.31	334.993
TOI-4145 A b	TOI-4145 A	4.0664428	0.04823	13.305	1.187	136.66622	0.43		5281	0.86	0.86	0.91	205.387
TOI-4153 b	TOI-4153	4.6174141	0.06311	16.119	1.438	365.50267	1.15		6860	1.6	1.6	1.57	423.84
TOI-4184 b	TOI-4184	4.9019804	0.0336	2.43	0.217			M5.5+/-0.5	3238	0.23	0.23	0.21	69.3649
TOI-4201 b	TOI-4201	3.5819194	0.04	13.675	1.22	788.21445	2.48	M1.0+/-0.5	3794	0.63	0.63	0.61	189.011
TOI-421 b	TOI-421	5.19672	0.056	2.68	0.239	7.17	0.02256	G9 V	5325	0.87	0.87	0.85	74.7969
TOI-421 c	TOI-421	16.06819	0.1189	5.09	0.454	16.42	0.05166	G9 V	5325	0.87	0.87	0.85	74.7969
TOI-4308 b	TOI-4308	9.151201	0.0113	2.419	0.216				5243	0.79	0.79	0.9	108.932
TOI-431 b	TOI-431	0.490047	0.052	1.28	0.114	3.07	0.00966		4850	0.73	0.73	0.78	32.5686
TOI-431 c	TOI-431	4.8494							4850	0.73	0.73	0.78	32.5686
TOI-431 d	TOI-431	12.46103	0.098	3.29	0.294	9.9	0.03115		4850	0.73	0.73	0.78	32.5686
TOI-4329 b	TOI-4329	2.9223		16.813	1.5	143.02278	0.45		6000	2.31	2.31	1.54	717.128
TOI-4336 A b	TOI-4336 A	16.336334	0.0872	2.12	0.189			M3.5+/-0.5	3298	0.33	0.33	0.33	22.5455
TOI-4342 b	TOI-4342	5.5382498	0.05251	2.266	0.202			M0 V	3901	0.6	0.6	0.63	61.5204
TOI-4342 c	TOI-4342	10.688716	0.0814	2.415	0.215			M0 V	3901	0.6	0.6	0.63	61.5204
TOI-4377 b	TOI-4377	4.378081	0.0579	15.11	1.348	304.16179	0.957		4974	3.52	3.52	1.36	455.82
TOI-4406 b	TOI-4406	30.08364	0.201	11.209	1	95.34852	0.3		6219	1.29	1.29	1.19	261.834
TOI-4438 b	TOI-4438	7.44628	0.0534	2.52	0.225	5.4	0.01699	M3.5 V	3422	0.37	0.37	0.37	30.0346
TOI-444 b	TOI-444	17.9636	0.133	2.77	0.247			K1/2 V	5225	0.78	0.78	0.96	57.3973
TOI-4443.01	TOI-4443	1.8498567	0.029963	1.719	0.153				5834	1.02	1.02	1.05	56.0739
TOI-4463 A b	TOI-4463 A	2.8807198	0.04036	13.26	1.183	252.35576	0.794		5640	1.06	1.06	1.06	173.117
TOI-4479 b	TOI-4479	1.1589	0.0164	2.82	0.252			M3.0	3400	0.45	0.45	0.45	80.6532
TOI-4495.01	TOI-4495	5.1830043	0.061733	3.63	0.324				6156	1.34	1.34	1.17	132.214
TOI-451 b	TOI-451	1.858703	0.0283	1.91	0.17				5550	0.88	0.88	0.95	123.739
TOI-451 c	TOI-451	9.192522	0.0823	3.1	0.277				5550	0.88	0.88	0.95	123.739
TOI-451 d	TOI-451	16.364988	0.1208	4.07	0.363				5550	0.88	0.88	0.95	123.739
TOI-4515 b	TOI-4515	15.266446	0.118	12.17	1.086	637	2.005	G8/G9	5433	0.86	0.86	0.95	193.499
TOI-4527.01	TOI-4527	0.3994445	0.008332	0.909	0.081				3702	0.49	0.49	0.48	18.1037

(*Continued*)

(Continued)

pl_name	hostname	pl_orbper	pl_orbsmax	pl_rade	pl_radj	pl_masse	pl_massj	st_spectype	st_teff	st_rad	st_mass	sy_dist
TOI-4551 b	TOI-4551	9.95581	0.0994	11.859	1.058	473.56433	1.49		4896	3.55	1.31	215.986
TOI-4559 b	TOI-4559	3.965991	0.0359	1.415	0.126				3558	0.37	0.39	31.6354
TOI-4562 b	TOI-4562	225.11781	0.768	12.53	1.118	732	2.3	F7 V	6096	1.15	1.19	339.605
TOI-4582 b	TOI-4582	31.034		10.536	0.94	168.44906	0.53		5190	2.5	1.34	383.846
TOI-4600 b	TOI-4600	82.6869	0.349	6.8	0.607	959.84179	3.02	K V	5170	0.81	0.89	216.056
TOI-4600 c	TOI-4600	482.8191	1.152	9.42	0.841	2946.26933	9.27	K V	5170	0.81	0.89	216.056
TOI-4602.01	TOI-4602	3.9812091	0.050997	2.55	0.227				6011.9	1.16	1.12	63.0637
TOI-4603 b	TOI-4603	7.24599	0.0888	11.68	1.042	4096.80817	12.89	F	6264	2.74	1.76	225.643
TOI-4633 c	TOI-4633	271.9445	0.847	3.2	0.285	123	0.387	G V	5800	1.05	1.1	95.2016
TOI-4641 b	TOI-4641	22.09341	0.173	8.183	0.73	1229.99594	3.87		6560	1.72	1.41	86.3501
TOI-470 b	TOI-470	12.19148	0.119	4.34	0.387			G	5190	0.83	0.87	130.46
TOI-4791 b	TOI-4791	4.28088	0.0555	12.442	1.11	734.18362	2.31		6058	1.41	1.24	328.802
TOI-480 b	TOI-480	6.866196	0.077	2.854	0.255	20.8	0.06544		6174	1.49	1.28	54.5306
TOI-481 b	TOI-481	10.33111	0.097	11.097	0.99	486.2799	1.53	G	5735	1.66	1.14	179.366
TOI-4860 b	TOI-4860	1.5227591	0.01808	8.7	0.77	86.7	0.273	M3.5 V	3260	0.36	0.34	80.1423
TOI-500 b	TOI-500	0.548177	0.01189	1.166	0.104	1.42	0.00447	K6 V	4440	0.68	0.74	47.3924
TOI-5076 b	TOI-5076	23.445		3.2	0.285	16	0.05034	K2 V	5070	0.78	0.8	82.857
TOI-5082.01	TOI-5082	4.240122	0.051205	2.548	0.227				5670	0.93	1	43.1406
TOI-5126 b	TOI-5126	5.4588385	0.06519	4.74	0.423			F V	6150	1.24	1.24	160.414
TOI-5126 c	TOI-5126	17.8999	0.1439	3.86	0.344			F V	6150	1.24	1.24	160.414
TOI-5153 b	TOI-5153	20.33003	0.158	11.882	1.06	1036.12061	3.26	F8 V	6300	1.4	1.24	390.484
TOI-5174 b	TOI-5174	12.214286		5.351	0.477				5643.77	1.09	1	197.243
TOI-519 b	TOI-519	1.2652328	0.0159	11.545	1.03	147.15455	0.463	M V	3322	0.35	0.34	115.557
TOI-5205 b	TOI-5205	1.630757	0.0199	11.6	1.03	343	1.08	M4.0	3430	0.39	0.39	86.4493
TOI-5232 b	TOI-5232	4.0966692	0.05593	12.778	1.14	743.71847	2.34		6500	1.78	1.39	609.388
TOI-5238 b	TOI-5238	4.872171		5.22	0.466				5631	1.02	1	290.646
TOI-5293 A b	TOI-5293 A	2.930289	0.034	11.9	1.06	170.4	0.54	M3+/-1	3586	0.52	0.54	162.229
TOI-530 b	TOI-530	6.387597	0.052	9.303	0.83	127.13136	0.4	M0.5 V	3659	0.54	0.53	148.762
TOI-5301 b	TOI-5301	5.858858	0.073	13.092	1.168	1182.32167	3.72		6240	2.19	1.48	629.161
TOI-532 b	TOI-532	2.3266508	0.0296	5.82	0.519	61.5	0.1935		3927	0.61	0.64	135.047
TOI-5344 b	TOI-5344	3.792622	0.04041	10.604	0.946	130.9453	0.412		3747	0.59	0.61	138.367

TOI-5388.01	TOI-5388	2.594675	0.024336	1.894	0.169				3495	0.31	0.29	18.5226
TOI-5398 b	TOI-5398	10.590547	0.098	10.3	0.919	58.7	0.18469	G V	6000	1.05	1.15	130.855
TOI-5398 c	TOI-5398	4.77271	0.057	3.52	0.314	11.8	0.03713	G V	6000	1.05	1.15	130.855
TOI-540 b	TOI-540	1.2391491	0.01223	0.903	0.081			M V	3216	0.19	0.16	14.0022
TOI-544 b	TOI-544	1.548352	0.0225	2.018	0.18	2.89	0.00909		4169	0.62	0.63	41.1166
TOI-554 b	HD 25463	7.0491423	0.076	2.545	0.227	8.7	0.02737		6353	1.42	1.25	45.6181
TOI-554 c	HD 25463	3.04405	0.0443	1.364	0.122	4.3	0.01353		6353	1.42	1.25	45.6181
TOI-5542 b	TOI-5542	75.12375	0.332	11.31	1.009	419.5335	1.32	G3 V	5700	1.06	0.89	348.849
TOI-558 b	TOI-558	14.574071	0.1291	12.173	1.086	1147.36055	3.61	F V	6466	1.5	1.35	402.63
TOI-559 b	TOI-559	6.9839095	0.0723	12.229	1.091	1910.14873	6.01	G V	5925	1.23	1.03	233.266
TOI-561 b	TOI-561	0.4465688	0.0106	1.425	0.127	2	0.00629	G9 V	5372	0.84	0.81	85.799
TOI-561 c	TOI-561	10.778831	0.0884	2.91	0.26	5.39	0.01696	G9 V	5372	0.84	0.81	85.799
TOI-561 d	TOI-561	25.7124	0.158	2.82	0.252	13.2	0.04153	G9 V	5372	0.84	0.81	85.799
TOI-561 e	TOI-561	77.03	0.328	2.55	0.227	12.6	0.03964	G9 V	5372	0.84	0.81	85.799
TOI-564 b	TOI-564	1.651144	0.02734	11.433	1.02	464.98529	1.463	G7 V	5640	1.09	1	199.581
TOI-5678 b	TOI-5678	47.73022	0.249	4.91	0.438	20	0.06293		5485	0.94	0.91	165.393
TOI-5704 b	TOI-5704	3.771116		3.227	0.288			M3.5 +/- 1.0	4590	0.76	0.73	89.637
TOI-5720 b	TOI-5720	1.4344555	0.018288	1.09	0.097	4.3	0.01353		3325	0.38	0.38	35.8256
TOI-5799 b	TOI-5799	4.164753	0.0352	1.625	0.145				3514	0.33	0.34	27.8123
TOI-5803 b	TOI-5803	5.38305		3.273	0.292				5134	0.76	0.87	84.0126
TOI-6008 b	TOI-6008	0.8574347	0.010777	1.03	0.092	4	0.01259	M5.0	3075	0.24	0.23	23.1008
TOI-6086 b	TOI-6086	1.3888725	0.01541	1.18	0.105			M3.0	3200	0.26	0.25	31.4566
TOI-615 b	TOI-615	4.6615983	0.0678	18.977	1.693	138.25536	0.435	F2 V	6850	1.73	1.45	360.408
TOI-620 b	TOI-620	5.0988179	0.04825	3.76	0.335			M2.5 V	3708	0.55	0.58	33.0231
TOI-622 b	TOI-622	6.402513	0.0708	9.236	0.824	96.30201	0.303	F6 V	6400	1.42	1.31	123.297
TOI-628 b	TOI-628	3.4095675	0.0486	11.882	1.06	2011.8639	6.33		6250	1.34	1.31	178.68
TOI-640 b	TOI-640	5.0037775	0.06608	19.851	1.771	279.6904	0.88		6460	2.08	1.54	341.997
TOI-654.01	TOI-654	1.5275657	0.02102	2.371	0.211				3433	0.43	0.53	57.7553
TOI-663 b	TOI-663	2.598902	0.0295	2.27	0.203	4.45	0.014	M1 V	3681	0.51	0.51	64.2308
TOI-663 c	TOI-663	4.6955507	0.0437	2.26	0.202	3.65	0.01148	M1 V	3681	0.51	0.51	64.2308
TOI-663 d	TOI-663	7.1027268	0.0576	1.92	0.171	5.2	0.01636	M1 V	3681	0.51	0.51	64.2308
TOI-669 b	TOI-669	3.945152	0.048	2.586	0.231	10	0.03146		5597	0.99	0.92	142.867

(Continued)

(Continued)

pl_name	hostname	pl_orbper	pl_orbsmax	pl_rade	pl_radj	pl_masse	pl_massj	st_spectype	st_teff	st_rad	st_mass	sy_dist
TOI-672 b	TOI-672	3.633575		5.26	0.469	23.6	0.07425		3765	0.54	0.54	66.9071
TOI-674 b	TOI-674	1.977143	0.025	5.25	0.468			M2 V	3514	0.42	0.42	46.0851
TOI-677 b	TOI-677	11.2366	0.1038	13.115	1.17	392.83788	1.236	F8	6295	1.28	1.18	141.812
TOI-700 b	TOI-700	9.977219	0.0677	0.914	0.082			M2.5 V	3459	0.42	0.41	31.1265
TOI-700 c	TOI-700	16.051137	0.0929	2.6	0.232			M2.5 V	3459	0.42	0.41	31.1265
TOI-700 d	TOI-700	37.42396	0.1633	1.073	0.096			M2.5 V	3459	0.42	0.41	31.1265
TOI-700 e	TOI-700	27.80978	0.134	0.953	0.085			M2.5 V	3459	0.42	0.41	31.1265
TOI-712 b	TOI-712	9.531361	0.07928	2.049	0.183			K4.5 V	4622	0.67	0.73	58.6212
TOI-712 c	TOI-712	51.69906	0.2447	2.701	0.241			K4.5 V	4622	0.67	0.73	58.6212
TOI-712 d	TOI-712	84.8396	0.3405	2.474	0.221			K4.5 V	4622	0.67	0.73	58.6212
TOI-715 b	TOI-715	19.288004	0.083	1.55	0.138			M4	3075	0.24	0.23	42.4048
TOI-733 b	TOI-733	4.884765	0.0618	1.992	0.178	5.72	0.018	G	5585	0.95	0.96	75.2092
TOI-762 A b	TOI-762 A	3.4716826	0.03418	8.339	0.744	79.77493	0.251		3266	0.42	0.44	98.4458
TOI-763 b	TOI-763	5.6057	0.06	2.28	0.203	9.79	0.0308	G	5450	0.9	0.92	95.1274
TOI-763 c	TOI-763	12.2737	0.1011	2.63	0.235	9.32	0.02932	G	5450	0.9	0.92	95.1274
TOI-771 b	TOI-771	2.326021	0.0207	1.422	0.127				3201	0.24	0.22	25.2788
TOI-776 b	TOI-776	8.24662	0.0653	1.798	0.16	5	0.01573		3725	0.55	0.54	27.1701
TOI-776 c	TOI-776	15.665323	0.1001	2.047	0.183	6.9	0.02171		3725	0.55	0.54	27.1701
TOI-778 b	TOI-778	4.633611	0.06	15.4	1.37	878	2.8	F3 V	6643	1.71	1.4	161.743
TOI-782 b	TOI-782	8.0240015	0.0578	2.734	0.244	19.1	0.0601		3370	0.41	0.4	52.5122
TOI-784 b	HD 307842	2.7970365	0.038	1.93	0.172	9.67	0.03043	G	5558	0.91	0.91	64.5997
TOI-813 b	TOI-813	83.8911	0.423	6.71	0.599			G0 IV	5907	1.94	1.32	263.145
TOI-815 b	TOI-815	11.197259	0.0903	2.94	0.262	7.6	0.02391	K3 V	4869	0.77	0.78	59.7117
TOI-815 c	TOI-815	34.976145	0.193	2.62	0.234	23.5	0.07394	K3 V	4869	0.77	0.78	59.7117
TOI-824 b	TOI-824	1.392978	0.02177	2.926	0.261	18.467	0.0581	K4 V	4600	0.69	0.71	63.9262
TOI-833 b	TOI-833	1.0418777	0.0171	1.27	0.113				3920	0.6	0.61	41.7148
TOI-836 b	TOI-836	3.81673	0.0422	1.704	0.152	4.53	0.01425	K V	4552	0.67	0.68	27.5024
TOI-836 c	TOI-836	8.59545	0.075	2.587	0.231	9.6	0.0302	K V	4552	0.67	0.68	27.5024
TOI-837 b	TOI-837	8.3248762		8.631	0.77			G0/F9 V	6047	1.02	1.12	142.488
TOI-849 b	TOI-849	0.765548	0.0155	3.64	0.325	41.8	0.13152		5257	0.97	0.97	225.734
TOI-858 B b	TOI-858 B	3.2797178	0.04435	14.067	1.255	349.61125	1.1	G0	5842	1.31	1.08	250.489

TOI-871 b	TOI-871	14.362565	0.1054	1.664	0.148				4929	0.72	0.76	68.0473
TOI-880.02	TOI-880	2.5735943	0.034191	2.778	0.248				4935	0.82	0.81	60.6679
TOI-892 b	TOI-892	10.62656	0.092	11.994	1.07	301.9385	0.95	F	6261	1.39	1.28	340.543
TOI-904 b	TOI-904	10.8772	0.056	2.426	0.216				3770.2	0.53	0.56	46.0891
TOI-904 c	TOI-904	83.9997	0.312	2.167	0.193				3770.2	0.53	0.56	46.0891
TOI-905 b	TOI-905	3.739494	0.04666	13.126	1.171	211.99261	0.667		5570	0.92	0.97	158.649
TOI-907.01	TOI-907	4.5849129	0.055225	9.616	0.857				6272	2.06	1.07	312.529
TOI-908 b	TOI-908	3.183792	0.041657	3.186	0.284	16.137	0.05077	G V	5626	1.03	0.95	175.748
TOI-913 b	TOI-913	11.098644		2.453	0.219				4969	0.73	0.82	65.1769
TOI-942 b	TOI-942	4.32421	0.04866	3.89	0.347					0.89	0.82	152.601
TOI-942 c	TOI-942	10.156272	0.08598	4.67	0.417					0.89	0.82	152.601
TOI-954 b	TOI-954	3.6849729	0.04963	9.55	0.852	55.30242	0.174	K V	5710	1.89	1.2	238.768
TOI-969 b	TOI-969	1.8237305	0.02636	2.765	0.247	9.1	0.02863		4435	0.67	0.73	77.2554
WASP-126 c	WASP-126	7.63				64.2	0.20199		5800	1.27	1.12	216.526
WASP-132 c	WASP-132	1.011534	0.0182	1.85	0.165	37.35	0.11752	K4	4714	0.75	0.78	122.91
WASP-18 c	WASP-18	2.1558	0.035	1.95	0.174	55.2	0.17368		6400	1.23	1.22	123.483
WASP-84 c	WASP-84	1.4468849	0.02359	10.4	0.928	15.2	0.048	K0		0.77	0.85	100.588
WD 1856 +534 b	WD 1856+534	1.4079405	0.0204	1.24	0.111	4386.054	13.8		4710	0.01	0.52	24.7359
Wolf 327 b	Wolf 327	0.5734745	0.01			2.53	0.00796	M2.5 V	3542	0.41	0.41	28.5321
pi Men c	HD 39091	6.26791	0.069								1.1	18.2702

www.ingramcontent.com/pod-product-compliance
Lightning Source LLC
Chambersburg PA
CBHW082105210326
41599CB00033B/6596